MATHEMATICAL GEOGRAPHY

INDEX

MATHEMATICAL GEOGRAPHY

Willis E. Johnson

Kaveri Books
New Delhi-110002

First Published in 2012
ISBN 978-81-7479-124-5

Published by
Kaveri Books
4832/24, Ansari Road
New Delhi - 110 002 (India)
Tel.: 011-2328 8140, 2324 5799,
Fax: 011-2328 8140
E-mail: kaveribooks@vsnl.com,
kaveribooks@gmail.com
Website: www.kaveribooks.com

Laser Typesetting by : Deep Graphics, Delhi

Printed at : Salasar Imaging Systems, Delhi-110 052

PRINTED IN INDIA

PREFACE

In the greatly awakened interest in the common-school subjects during recent years, geography has received a large share. The establishment of chairs of geography in some of our greatest universities, the giving of college courses in physiography, meteorology, and commerce, and the general extension of geography courses in normal schools, academies, and high schools, may be cited as evidence of this growing appreciation of the importance of the subject.

While physiographic processes and resulting land forms occupy a large place in geographical control, the earth in its simple mathematical aspects should be better understood than it generally is, and mathematical geography deserves a larger place in the literature of the subject than the few pages generally given to it in our physical geographies and elementary astronomies. It is generally conceded that the mathematical portion of geography is the most difficult, the most poorly taught and least understood, and that students require the most help in understanding it. The subject-matter of mathematical geography is scattered about in many works, and no one book treats the subject with any degree of thoroughness, or even makes a pretense at doing so. It is with the view of meeting the need for such a volume that this work has been undertaken.

Although designed for use in secondary schools and for teachers' preparation, much material herein organized may be used in the upper grades of the elementary school. The subject has not been presented from the point of view of a little child, but an attempt has been made to keep its scope within the attainments of a student in a normal school, academy, or high school. If a very short course in mathematical geography is given, or if students are relatively advanced, much of the subject-matter may be omitted or given as special reports.

To the student or teacher who finds some portions too difficult, it is suggested that the discussions which seem obscure at first reading are often made clear by additional explanation given further on in the book. Usually the second study of a topic which seems too difficult should be deferred until the entire chapter has been read over carefully.

The experimental work which is suggested is given for the purpose of making the principles studied concrete and vivid. The measure of the educational value of a laboratory exercise in a school of secondary grade is not found in the academic results obtained, but in the attainment of a conception of a process. The student's determination of latitude, for example, may not be of much value if its worth is estimated in terms of facts obtained, but the forming of the conception of the process is a result of inestimable educational value. Much time may be wasted, however, if the student is required to rediscover the facts and laws of nature which are often so simple that to see is to accept and understand.

To President George W. Nash of the Northern Normal and Industrial School, who carefully read the entire manuscript and proof, and to whose thorough training, clear insight, and kindly interest the author is under deep obligations, especial credit is gratefully accorded. While the author has not availed himself of the direct assistance of his sometime teacher, Professor Frank E. Mitchell, Head of the Department of Geography and Geology of the State Normal School at Oshkosh, Wisconsin, he wishes formally to acknowledge his obligation to him for an abiding interest in the subject. For the critical examination of portions of the manuscript bearing upon fields in which they are acknowledged authorities, grateful acknowledgment is extended to Professor Francis P. Leavenworth, Head of the Department of Astronomy of the University of Minnesota; to Lieutenant-Commander E. E. Hayden, Head of the Department of Chronometers and Time Service of the United States Naval Observatory, Washington; to President F. W. McNair of the Michigan College of Mines; to Professor Cleveland Abbe of the United States Weather Bureau; to President Robert S. Woodward of the Carnegie Institution of Washington; to Professor T. C. Chamberlin, Head of the Department of Geology of the University of Chicago; and to Professor Charles R. Dryer, Head of the Department of Geography of the State Normal School at Terre Haute, Indiana.

CONTENTS

CHAPTER-I

INTRODUCTION

OBSERVATIONS AND EXPERIMENTS

Observations of the Stars. On the first clear evening, observe the "Big Dipper"* and the polestar. In September and in December, early in the evening, they will be nearly in the positions represented in Figure 1. Where is the Big Dipper later in the evening? Find out by observations.

Learn readily to pick out Cassiopeia's Chair and the Little Dipper. Observe their apparent motions also. Notice the positions of stars in different portions of the sky and observe where they are later in the evening. Do the stars around the polestar remain in the same position in relation to each other, — the Big Dipper always like a dipper, Cassiopeia's Chair always like a chair, and both always on

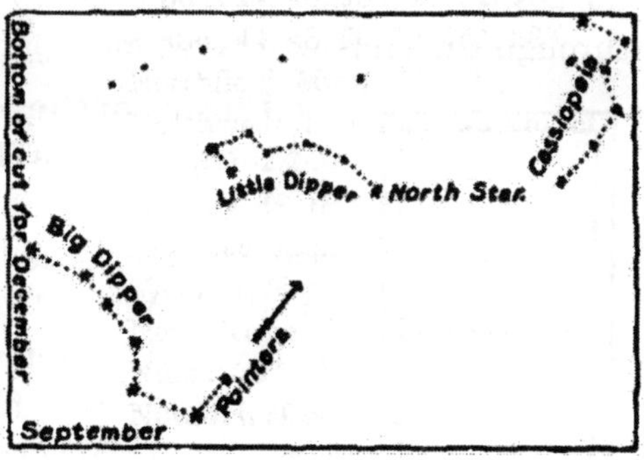

Fig. 1

*In Ursa Major, commonly called the "Plow," "The Great Wagon," or "Charles's Wagon" in England, Norway, Germany, and other countries.

opposite sides of the polestar? In what sense may they be called "fixed" stars (see pp. 109, 265)?

Make a sketch of the Big Dipper and the polestar, recording the date and time of observation. Preserve your sketch for future reference, marking it Exhibit 1. A month or so later, sketch again at the same time of night, using the same sheet of paper with a common polestar for both sketches. In making your sketches be careful to get the angle formed by a line through the "pointers" and the polestar with a perpendicular to the horizon. This angle can be formed by observing the side of a building and the pointer line. It can be measured more accurately in the fall months with a pair of dividers having straight edges, by placing one outer edge next to the perpendicular side of a north window and opening the dividers until the other outside edge is parallel to the pointer line (see Fig. 2). Now lay the dividers on a sheet of paper and mark the angle thus formed, representing the positions of stars with asterisks. Two penny rulers pinned through the ends will serve for a pair of dividers.

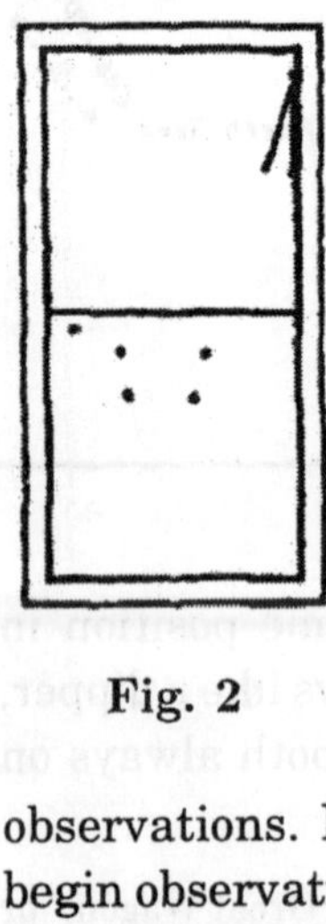

Fig. 2

Phases of the Moon. Note the position of the moon in the sky on successive nights at the same hour. Where does the moon rise? Does it rise at the same time from day to day? When the full moon may be observed at sunset, where is it? At sunrise? When there is a full moon at midnight, where is it? Assume it is sunset and the moon is high in the sky, how much of the lighted part can be seen?

Answers to the foregoing questions should be based upon first-hand observations. If the questions cannot easily be answered, begin observations at the first opportunity. Perhaps the best time to begin is when both sun and moon may be seen above the horizon. At each observation notice the position of the sun and of the moon, the portion of the lighted part which is turned toward the earth, and bear in mind the simple fact that *the moon always shows a lighted half to the sun*. If the moon is rising when the sun is setting, or the sun is rising when the moon is setting, the observer must be standing directly between them, or approximately so. With the sun at your back in the east and facing the moon in the west, you see the moon as though you were at the sun. How much of the lighted part of the moon is then seen? By far the best apparatus for illustrating the phases of the moon is the sun and moon themselves, especially when both are observed above the horizon.

The Noon Shadow. Some time early in the term from a convenient south window, measure upon the floor the length of the shadow when it is shortest during the day. Record the measurement and the date and time of day. Repeat the measurement each week. Mark this Exhibit 2.

On a south-facing window sill, strike a north-south line (methods for doing this are discussed on pp. 59, 130). Erect at the south end of this line a perpendicular board, say six inches wide and two feet long, with the edge next the north-south line. True it with a plumb line; one made with a bullet and a thread will do. This should be so placed that the shadow from the edge of the board may be recorded on the window sill from 11 o'clock, A.M., until 1 o'clock, p.m. (see Fig. 3).

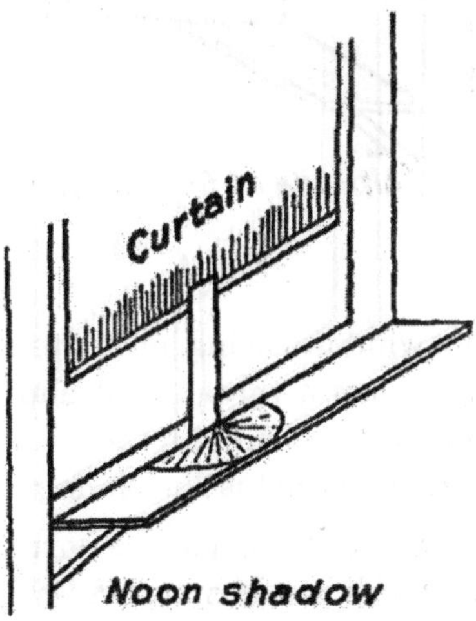

Fig. 3

Carefully cut from cardboard a semicircle and mark the degrees, beginning with the middle radius as zero. Fasten this upon the window sill with the zero meridian coinciding with the north-south line. Note accurately the clock time when the shadow from the perpendicular board crosses the line, also where the shadow is at twelve o'clock. Record these facts with the date and preserve as Exhibit 3. Continue the observations every few days.

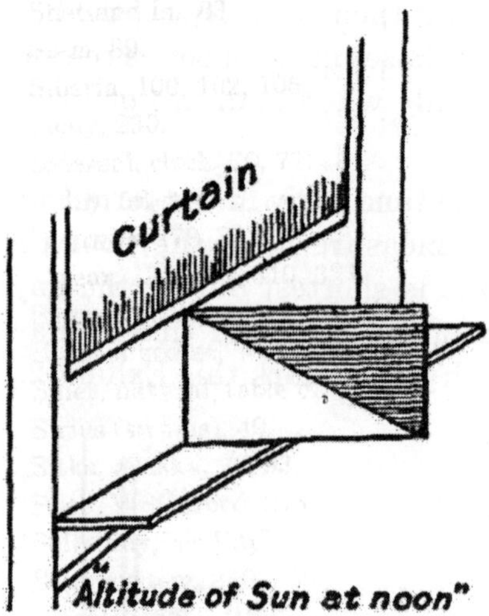

Fig. 4

The Sun's Meridian Altitude. When the shadow is due north, carefully measure the angle formed by the shadow and a level line. The simplest way is to draw the window shade down to the top of a sheet of cardboard placed very nearly north and south with the bottom level and then draw the shadow line, the lower acute angle being the one sought (see Fig. 4). Another way is to drive a pin in the side of the window casing, or in the edge of the vertical board (Fig. 3); fasten a thread to it and connect the other end of the thread to a point on the sill where the shadow falls. A still better method is shown on p. 172.

Since the shadow is north, the sun is as high in the sky as it will get during the day, and the angle thus measured gives the highest altitude of the sun for the day. Record the measurement of the angle with the date as Exhibit 4. Continue these records from week to week, especially noting the angle on one of the following dates: March 21, June 22, September 23, December 22. This angle on March 21 or September 23, if subtracted from 90°, will equal the latitude* of the observer.

A FEW TERMS EXPLAINED

Centrifugal Force. The literal meaning of the word suggests its current meaning. It comes from the Latin *centrum*, center; and *fugere*, to flee. A centrifugal force is one directed away from a center. When a stone is whirled at the end of a string, the pull which the stone gives the string is called centrifugal force. Because of the inertia of the stone, the whirling motion given to it by the arm tends to make it fly off in a straight line (Fig. 5), — and this it will do if the string breaks. The measure of the centrifugal force is the tension on the string. If the string be fastened at the end of a spring scale and the stone whirled, the scale will show the amount of the centrifugal force which is given the stone by the arm that whirls it. The amount of this force† *(C)* varies with the mass of the body (*m*), its velocity (*v*), and the radius

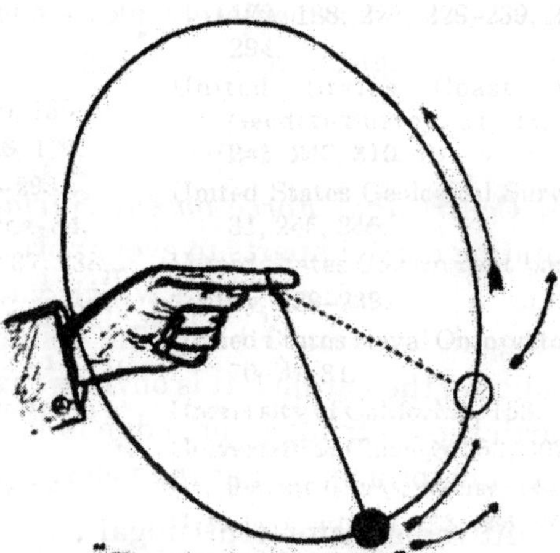

Fig. 5

*This is explained on p. 170.

† On the use of symbols, such as C for centrifugal force, f for latitude, etc., see Appendix, p. 306.

of the circle (r) in which it moves, in the following ratio:

$$C = \frac{mv^2}{r}.$$

The instant that the speed becomes such that the available strength of the string is less than the value of $\frac{mv^2}{r}$, however slightly, the stone will cease to follow the curve and will immediately take a motion at a uniform speed in the straight line with which its motion happened to coincide at that instant (a tangent to the circle at the point reached at that moment).

Centrifugal Force on the Surface of the Earth. The rotating earth imparts to every portion of it, save along the axis, a centrifugal force which varies according to the foregoing formula, r being the distance to the axis, or the radius of the parallel. It is obvious that on the surface of the earth the centrifugal force due to its rotation is greatest at the equator and zero at the poles.

At the equator centrifugal force (C) amounts to about $\frac{1}{289}$ that of the earth's attraction (g), and thus an object there which weighs 288 pounds is lightened just one pound by centrifugal force, that is, it would weigh 289 pounds were the earth at rest. At latitude 30°, $C = \frac{g}{385}$ = (that is, centrifugal force is $\frac{1}{385}$ the force of the earth's attraction); at 45°, $C = \frac{g}{578}$; at 60°, $C = \frac{g}{1156}$.

For any latitude the "lightening effect" of centrifugal force due to the earth's rotation equals $\frac{g}{289}$ times the square of the cosine of the latitude $(C = \frac{g}{289} \times \cos^2 \phi)$. By referring to the table of cosines in the Appendix, the student can easily calculate the "lightening" influence of centrifugal force at his own latitude.

For example, say the latitude of the observer is 40°.

$$\text{Cosine } 40^\circ = .7660. \qquad \frac{g}{289} \times .7660^2 = \frac{g}{492}.$$

Thus the earth's attraction for an object on its surface at latitude 40° is 492 times as great as centrifugal force there, and an object weighing 491 pounds at that latitude would weigh one pound more were the earth at rest.*

Centripetal Force. A centripetal (*centrum*, center; *petere*, to seek) force is one directed toward a center, that is, at right angles to the direction of motion of a body. To distinguish between centrifugal force and centripetal force, the student should realize that forces never occur singly but only in pairs and that in any force action there are always *two bodies* concerned. Name them *A* and *B*. Suppose *A* pushes or pulls *B* with a certain strength. This cannot occur except *B* pushes or pulls *A* by the same amount and in the opposite direction. This is only a simple way of stating Newton's third law that to every action (*A* on *B*) there corresponds an equal and opposite reaction (*B* on *A*).

Centrifugal force is the *reaction* of the body against the centripetal force which holds it in a curved path, and it must always exactly equal the centripetal force. In the case of a stone whirled at the end of a string, the necessary force which the string exerts on the stone to keep it in a curved path is centripetal force, and the reaction of the stone upon the string is centrifugal force.

The formulas for centripetal force are exactly the same as those for centrifugal force. Owing to the rotation of the earth, a body at the equator describes a circle with uniform speed. The attraction of the earth supplies the centripetal force required to hold it in the circle. The earth's attraction

*These calculations are based upon a spherical earth and make no allowances for the oblateness.

is greatly in excess of that which is required, being, in fact, 289 times the amount needed. *The centripetal force in this case is that portion of the attraction which is used to hold the object in the circular course.* The excess is what we call the weight of the body or the force of gravity.

If, therefore, a spring balance suspending a body at the equator shows 288 pounds, we infer that the earth really pulls it with a force of 289 pounds, but one pound of this pull is expended in changing the direction of the motion of the body, that is, the value of centripetal force is one pound. The body pulls the earth to the same extent, that is, the centrifugal force is also one pound. At the poles neither centripetal nor centrifugal force is exerted upon bodies and hence the weight of a body there is the full measure of the attraction of the earth.

Gravitation. Gravitation is the all-pervasive force by virtue of which every particle of matter in the universe is constantly drawing toward itself every other particle of matter, however distant. The amount of this attractive force existing between two bodies depends upon (1) the amount of matter in them, and (2) the distance they are apart.

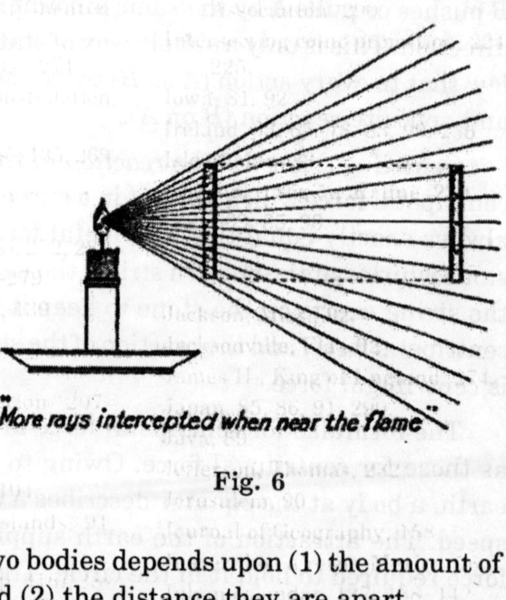

"More rays intercepted when near the flame"

Fig. 6

There are thus two laws of gravitation. The first law, the greater the mass, or amount of matter, the greater the attraction, is due to the fact that each particle of matter has its own independent attractive force, and the more there are of the particles, the greater is the combined attraction.

The Second Law Explained. In general terms the law is that the nearer an object is, the greater is its attractive force. Just as the heat and light of a flame are greater the nearer one gets to it (Fig. 6), because more rays are intercepted, so the nearer an object is, the greater is its attraction. The ratio of the increase of the power of gravitation as distance decreases, may be seen from Figures 7 and 8.

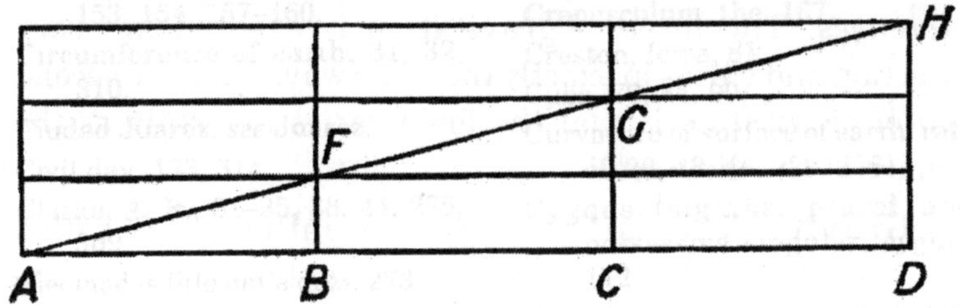

Fig. 7

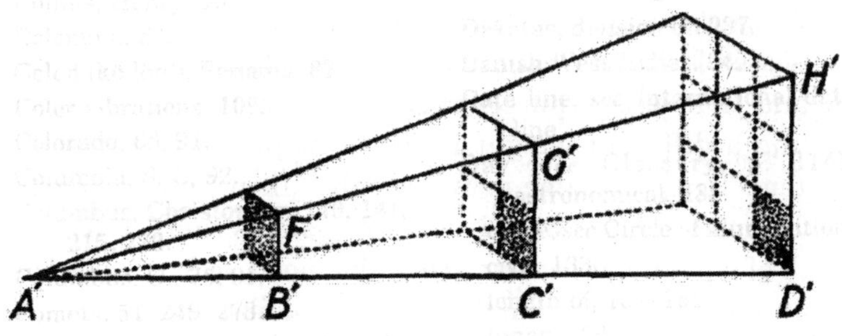

Fig. 8

Two lines, AD and AH (Fig. 7), are twice as far apart at C as at B because twice as far away; three times as far apart at D as at B because three times as far away, etc. Now light radiates out in every direction, so that light coming from point A'. (Fig. 8), when it reaches *B'* will be spread over the square of *B'F'*; at *C'*, on the square *C'G'*; at *D'* on the square *D'H'*, etc. *C'* being twice as far away from *A'* as *B'*, the side *C'G'* is twice that of *B'F'*, as we observed in Fig. 7, and its square is four times as great. Line *D'H'* is three times as far away, is three times as long, and its square is nine times as great. The light being spread over more space in the more distant objects, it will light up a given area less. The square at *B'* receives all the light within the four radii, the same square at *C'* receives one fourth of it, at *D'* one ninth, etc. The amount of light *decreases as the square of the distance increases.* The force of gravitation is exerted in every direction and varies in exactly the same way. Thus the second law of gravitation is that the force varies inversely as the square of the distance.

Gravity. The earth's attractive influence is called *gravity*. The weight of an object is simply the measure of the force of gravity. An object on or above the surface of the earth weighs less as it is moved away from the center of gravity.* It is difficult to realize that what we call the weight of an object is simply the excess of attraction which the earth possesses for it as compared with other forces acting upon it, and that it is a purely relative affair, the same object having a different weight in different places in the solar system. Thus the same pound-weight taken from the earth to the sun's surface would weigh 27 pounds there, only one sixth of a pound at the surface of the moon, over 2½ pounds on Jupiter, etc. If the earth were more dense, objects would weigh more on the surface. Thus if the earth retained its present size but contained as much matter as the sun has, the strongest man in the world could not lift a silver half dollar, for it would then weigh over five tons. A pendulum

*For a more accurate and detailed discussion of gravity, see p. 278.

clock would then tick 575 times as fast. On the other hand, if the earth were no denser than the sun, a half dollar would weigh only a trifle more than a dime now weighs, and a pendulum clock would tick only half as fast.

From the table on p. 266 giving the masses and distances of the sun, moon, and principal planets, many interesting problems involving the laws of gravitation may be suggested. To illustrate, let us take the problem "What would you weigh if you were on the moon?"

Weight on the Moon. The mass of the moon, that is, the amount of matter in it, is $\frac{1}{81}$ that of the earth. Were it the same size as the earth and had this mass, one pound on the earth would weigh a little less than one eightieth of a pound there. According to the first law of gravitation we have this proportion:

1. Earth's attraction : Moon's attraction : : $1 : \frac{1}{81}$.

But the radius of the moon is 1081 miles, only a little more than one fourth that of the earth. Since a person on the moon would be so much nearer the center of gravity than he is on the earth, he would weigh much more there than here if the moon had the same mass as the earth. According to the second law of gravitation we have this proportion:

2. Earth's attraction : Moon's attraction : : $\frac{1}{4000^2} : \frac{1}{1081^2}$.

We have then the two proportions:

1. Att. Earth : Att. Moon : : $1 : \frac{1}{81}$.
2. Att. Earth : Att. Moon $\frac{1}{4000^2} : \frac{1}{1081^2}$.

Combining these by multiplying, we get

Att. Earth : Att. Moon : : 6 : 1.

Therefore six pounds on the earth would weigh only one pound on the moon. Your weight, then, divided by six, represents what it would be on the moon. There you could jump six times as high—if you could live to jump at all on that cold and almost airless satellite (see p. 263).

The Sphere, Circle, and Ellipse. A *sphere* is a solid bounded by a curved surface all points of which are equally distant from a point within called the center.

A *circle* is a plane figure bounded by a curved line all points of which are equally distant from a point within called the center. In geography what we commonly call circles such as the equator, parallels, and meridians, are really only the circumferences of circles. Wherever used in this book, unless otherwise stated, the term circle refers to the circumference.

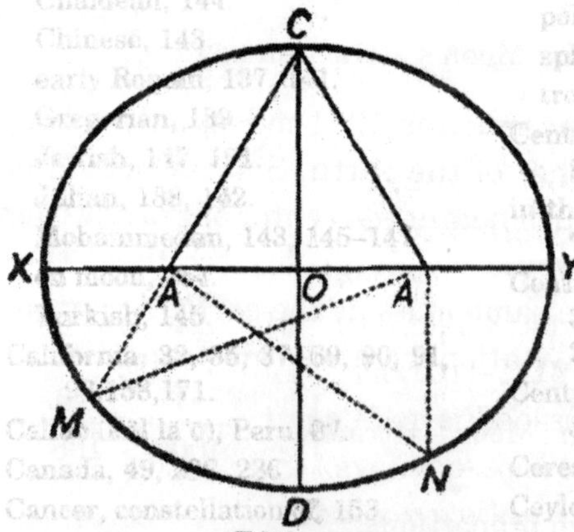

Ellipse.

A&A'. Foci. C D. Minor Axis. XY. Major Axis. A to A'. Focal Distance. AM+A'M=AN+A'N

Fig. 9

Every circle is conceived to be divided into 360 equal parts called degrees. The greater the size of the circle, the greater is the length of each degree. A *radius* of a circle or of a sphere is a straight line from the boundary to the center. Two radii forming a straight line constitute a *diameter*.

Circles on a sphere dividing it into two hemispheres are called great circles. Circles on a sphere dividing it into unequal parts are called small circles.

All great circles on the same sphere bisect each other, regardless of the angle at which they cross one another. That this may be clearly seen, with a globe before you test these two propositions:

a. A point 180° in any direction from one point in a great circle must lie in the same circle.

b. Two great circles on the same sphere must cross somewhere, and the point 180° from the one where they cross, lies in both of the circles, thus each great circle divides the other into two equal parts.

An angle is the difference in direction of two lines which, if extended, would meet. Angles are measured by using the meeting point as the center of a circle and finding the fraction of the circle, or number of degrees of the circle, included between the lines. It is well to practice estimating different angles and then to test the accuracy of the estimates by reference to a graduated quadrant or circle having the degrees marked.

An *ellipse* is a closed plane curve such that the sum of the distances from one point in it to two fixed points within, called foci, is equal to the sum of the distances from any other point in it to the foci. The ellipse is a conic section formed by cutting a right cone by a plane passing obliquely through its opposite sides (see Ellipse in Glossary).

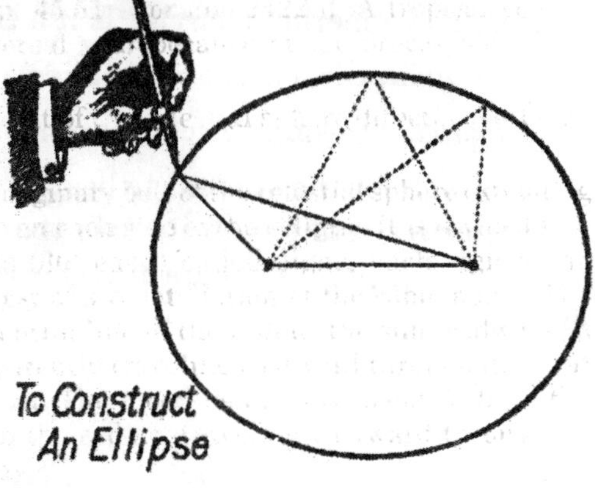

Fig. 10

To construct an ellipse, drive two pins at points for foci, say three inches apart. With a loop of non-elastic cord, say ten inches long, mark the boundary line as represented in Figure 10.

Orbit of the Earth. The orbit of the earth is an ellipse. To lay off an ellipse which shall quite correctly represent the shape of the earth's orbit, place pins one tenth of an inch apart and make a loop of string 12.2 inches long. This loop can easily be made by driving two pins 6.1 inches apart and tying a string looped around them.

Shape of the Earth. The earth is a spheroid, or a solid approaching a sphere (see Spheroid in Glossary). The diameter upon which it rotates is called the axis. The ends of the axis are its *poles*. Imaginary lines on the surface of the earth extending from pole to pole are called *meridians*.* While any number of meridians may be conceived of, we usually think of them as one degree apart. We say, for example, the ninetieth meridian, meaning the meridian ninety degrees from the prime or initial meridian. What kind of a circle is a meridian circle? Is it a true circle? Why?

The *equator* is a great circle midway between the poles.

Parallels are small circles parallel to the equator.

It is well for the student to bear in mind the fact that it is the earth's rotation on its axis that determines most of the foregoing facts. A sphere at rest would not have equator, meridians, etc.

*The term meridian is frequently used to designate a great circle passing through the poles. In this book such a circle is designated a *meridian circle*, since each meridian is numbered regardless of its opposite meridian.

CHAPTER-II

THE FORM OF THE EARTH

THE EARTH A SPHERE

Circumnavigation. The statements commonly given as proofs of the spherical form of the earth would often apply as well to a cylinder or an egg-shaped or a disk-shaped body. "People have sailed around it," "The shadow of the earth as seen in the eclipse of the moon is always circular," etc., do not in themselves prove that the earth is a sphere. They might be true if the earth were a cylinder or had the shape of an egg. "But men have sailed around it in different directions." So might they a lemon-shaped body. To make a complete proof, we must show that men have sailed around it in practically every direction and have found no appreciable difference in the distances in the different directions.

Earth's Shadow always Circular. The shadow of the earth as seen in the lunar eclipse is always circular. But a dollar, a lemon, an egg, or a cylinder may be so placed as always to cast a circular shadow. When in addition to this statement it is shown that the earth presents many different sides toward the sun during different eclipses of the moon and the shadow is always circular, we have a proof positive, for nothing but a sphere casts a circular shadow when in many different positions. The fact that eclipses of the moon are seen in different seasons and at different times of day is abundant proof that practically all sides of the earth are turned toward the sun during different eclipses.

Fig. 11. Ship's rigging distinct. Water hazy.

Almost Uniform Surface Gravity. An object has almost exactly the same weight in different parts of the earth (that is, on the surface), showing a practically common distance from different points on the earth's surface to the center of gravity. This is ascertained, not by carrying an object all over the earth and weighing it with a pair of spring scales (why not balances?); but by noting the time of the swing of the pendulum, for the rate of its swing varies according to the force of gravity.

Telescopic Observations. If we look through a telescope at a distant object over a level surface, such as a body of water, the lower part is hidden by the intervening curved surface. (Figs. 11, 12.) This has been observed in many different places, and the rate of curvature seems uniform everywhere and in every direction. Persons ascending in balloons or living on high elevations note the appreciably earlier time of sunrise or later time of sunset at the higher elevation.

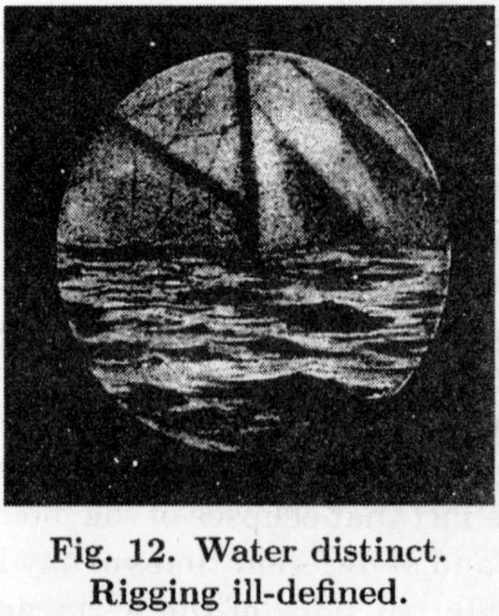

Fig. 12. Water distinct. Rigging ill-defined.

Shifting of Stars and Difference in Time. The proof which first demonstrated the curvature of the earth, and one which the student should clearly understand, is the disappearance of stars from the southern horizon and the rising higher of stars from the northern horizon to persons traveling north, and the sinking of northern stars and the rising of southern stars to south-bound travelers. After people had traveled far enough north and south to make an appreciable difference in the position of stars, they observed this apparent rising and sinking of the sky. Now two travelers, one going north and the other going south, will see the sky apparently elevated and depressed at the same time; that is, the portion of the sky that is rising for one will be sinking for the other. Since it is impossible that the stars be both rising and sinking at the same time, only one conclusion can follow,—the movement of the stars is apparent, and the path traveled north and south must be curved.

Owing to the rotation of the earth one sees the same stars in different positions in the sky east and west, so the proof just given simply shows that the earth is curved in a north and south direction. Only when timepieces were invented which could carry the time of one place to different portions of the earth could the apparent positions of the stars prove the curvature of the earth east and west. By means of the telegraph and telephone we have most excellent proof that the earth is curved east and west.

If the earth were at, when it is sunrise at Philadelphia it would be sunrise also at St. Louis and Denver. Sun rays extending to these places which are so near together as compared with the tremendous distance of the sun, over ninety millions of miles away, would be almost parallel on the earth and would strike these points at about the same angle. But we know from the many daily messages between these cities that sun time in Philadelphia is an hour later than it is in St. Louis and two hours later than in Denver.

When we know that the curvature of the earth north and south as observed by the general and practically uniform rising and sinking of the stars to north-bound and south-bound travelers is the same as the curvature east and west as shown by the difference in time of places east and west, we have an excellent proof that the earth is a sphere.

Actual Measurement. Actual measurement in many different places and in nearly every direction shows a practically uniform curvature in the different directions. In digging canals and laying watermains, an allowance must always be made for the curvature of the earth; also in surveying, as we shall notice more explicitly farther on.

A simple rule for finding the amount of curvature for any given distance is the following:

Square the number of miles representing the distance, and two thirds of this number represents in feet the departure from a straight line.

Suppose the distance is 1 mile. That number squared is 1, and two thirds of that number of feet is 8 inches. Thus an allowance of 8 inches must be made for 1 mile. If the distance is 2 miles, that number squared is 4, and two thirds of 4 feet is 2 feet, 8 inches. An object, then, 1 mile away sinks 8 inches below the level line, and at 2 miles it is below 2 feet, 8 inches.

To find the distance, the height from a level line being given, we have the converse of the foregoing rule:

Multiply the number representing the height in feet by 1½, and the square root of this product represents the number of miles distant the object is situated.

The following table is based upon the more accurate formula: Distance (Miles) $= 1:317\sqrt{\text{height (feet)}}$.

THE FORM OF THE EARTH

Ht. ft.	Dist. miles	Ht. ft.	Dist. miles	Ht. ft.	Dist. miles
1	1.32	50	9.31	170	17.17
2	1.86	55	9.77	180	17.67
3	2.28	60	10.20	190	18.15
4	2.63	65	10.62	200	18.63
5	2.94	70	11.02	300	22.81
6	3.23	75	11.40	400	26.34
7	3.48	80	11.78	500	29.45
8	3.73	85	12.14	600	32.26
9	3.95	90	12.49	700	34.84
10	4.16	95	12.84	800	37.25
15	5.10	100	13.17	900	39.51
20	5.89	110	13.81	1000	41.65
25	6.59	120	14.43	2000	58.90
30	7.21	130	15.02	3000	72.13
35	7.79	140	15.58	4000	83.30
40	8.33	150	16.13	5000	93.10
45	8.83	160	16.66	Mile	95.70

THE EARTH AN OBLATE SPHEROID

Richer's Discovery. In the year 1672 John Richer, the astronomer to the Royal Academy of Sciences of Paris, was sent by Louis XIV to the island of Cayenne to make certain astronomical observations. His Parisian clock had its pendulum, slightly over 39 inches long, regulated to beat seconds. Shortly after his arrival at Cayenne, he noticed that the clock was losing time, about two and a half minutes a day. Gravity, evidently, did not act with so much force near the equator as it did at Paris. The astronomer found it necessary to shorten the pendulum nearly a quarter of an inch to get it to swing fast enough.

Richer reported these interesting facts to his colleagues at Paris, and it aroused much discussion. At first it was thought that greater centrifugal force at the equator, counteracting the earth's attraction more there than elsewhere, was the explanation. The difference in the force of gravity, however, was soon discovered to be too great to be thus accounted for. The only other conclusion was that Cayenne must be farther from the center of gravity than Paris (see the discussion of Gravity, Appendix, p. 278; also Historical Sketch, pp. 272–274).

Repeated experiments show it to be a general fact that pendulums swing faster on the surface of the earth as one approaches the poles. Careful measurements of arcs of meridians prove beyond question that the earth is attened toward the poles, somewhat like an oblate spheroid. Further evidence is found in the fact that the sun and planets, so far as ascertained, show this same attening.

Cause of Oblateness. The cause of the oblateness is the rotation of the body, its attening e”ects being more marked in earlier plastic stages, as the earth and other planets are generally believed to have been at one time. The reason why rotation causes an equatorial bulging is not difficult to understand. Centrifugal force increases away from the poles toward the equator and gives a lifting or lightening inuence to portions on the surface. If the earth were a sphere, an object weighting 289 pounds at the poles would be lightened just one pound if carried to the swiftly rotating equator (see p. 279). The form given the earth by its rotation is called an oblate spheroid or an ellipsoid of rotation.

Amount of Oblateness. To represent a meridian circle accurately, we should represent the polar diameter about $\frac{1}{300}$ part shorter than the equatorial diameter. That this difference is not perceptible to the unaided eye will be apparent if the construction of such a figure is attempted, say ten inches in diameter in one direction and $\frac{1}{30}$ of an inch less in the opposite direction. The oblateness of Saturn is easily perceptible, being thirty times as great as that of the earth, or one tenth (see p. 258).Thus an ellipsoid nine inches in polar diameter (minor axis) and ten inches in equatorial diameter (major axis) would represent the form of that planet.

Although the oblateness of the earth seems slight when represented on a small scale and for most purposes may be ignored, it is nevertheless of vast importance in many problems in surveying, astronomy, and other subjects. Under the discussion of latitude it will be shown how this oblateness makes a difference in the lengths of degrees of latitude, and in the Appendix it is shown how this equatorial bulging shortens the length of the year and changes the inclination of the earth's axis (see Precession of the Equinoxes and Motions of the Earth's Axis).

Dimensions of the Spheroid. It is of very great importance in many ways that astronomers and surveyors know as exactly as possible the dimensions of the spheroid. Many men have made estimates based upon astronomical facts, pendulum experiments and careful surveys, as to the equatorial and polar diameters of the earth. Perhaps the most widely used is the one made by A. R. Clarke, for many years at the head of the English Ordnance Survey, known as the Clarke Spheroid of 1866.

CLARKE SPHEROID OF 1866.

A.	Equatorial diameter	7,926.614 miles
B.	Polar diameter	7,899.742 miles
	Oblateness $\frac{A-B}{A}$	$\frac{1}{195}$

It is upon this spheroid of reference that all of the work of the United States Geological Survey and of the United States Coast and Geodetic Survey is based, and upon which most of the dimensions given in this book are determined.

In 1878 Mr. Clarke made a recalculation, based upon additional information, and gave the following dimensions, though it is doubtful whether these approximations are any more nearly correct than those of 1866.

CLARKE SPHEROID OF 1878.

A. Equatorial diameter .... 7, 926.592 miles

B. Polar diameter .... 7, 899.580 miles

Oblateness $\frac{A-B}{A}$ .... $\frac{1}{293.46}$

Another standard spheroid of reference often referred to, and one used by the United States Governmental Surveys before 1880, when the Clarke spheroid was adopted, was calculated by the distinguished Prussian astronomer, F. H. Bessel, and is called the

BESSEL SPHEROID OF 1841.

A. Equatorial diameter .... 7, 925.446 miles

B. Polar diameter .... 7, 898.954 miles

Oblateness $\frac{A-B}{A}$ .... $\frac{1}{299.16}$

Many careful pendulum tests and a great amount of scientific triangulation surveys of long arcs of parallels and meridians within recent years have made available considerable data from which to determine the true dimensions of the spheroid. In 1900, the United States Coast and Geodetic Survey completed the measurement of an arc across the United States along the 39th parallel from Cape May, New Jersey, to Point Arena, California, through 48° 46' of longitude, or a distance of about 2, 625 miles. This is the most extensive piece of geodetic surveying ever undertaken by any nation and was so carefully done that the total amount of probable error does not amount to more than about eighty-five feet. A long arc has been surveyed diagonally from Calais, Maine, to New Orleans, Louisiana, through 15° 1' of latitude and 22° 47' of longitude, a distance of 1,623 miles. Another long arc will soon be completed along the 98th meridian across the United States. Many shorter arcs have also been surveyed in this country.

The English government undertook in 1899 the gigantic task of measuring the arc of a meridian extending the entire length of Africa, from Cape Town to Alexandria. This will be, when completed, 65° long, about half on each side of the equator, and will be of great value in determining the oblateness. Russia and Sweden have lately completed the measurement of an arc of 4° 30' on the island of Spitzbergen, which from its high latitude, 76° to 80° 30' N., makes it peculiarly valuable. Large arcs have been measured in India, Russia, France, and other countries, so that there are now available many times as much data from which the form and dimensions of the earth may be determined as Clarke or Bessel had.

The late Mr. Charles A. Schott, of the United States Coast and Geodetic Survey, in discussing the survey of the 39th parallel, with which he was closely identified, said:*

"Abundant additional means for improving the existing deductions concerning the earth's figure are now at hand, and it is perhaps not too much to expect that the International Geodetic Association may find it desirable in the near future to attempt a new combination of all available arc measures, especially since the two large arcs of the parallel, that between Ireland and Poland and that of the United States of America, cannot fail to have a paramount inuence in a new general discussion."

A spheroid is a solid nearly spherical. An oblate spheroid is one attened toward the poles of its axis of rotation. The earth is commonly spoken of as a sphere. It would be more

*In his Transcontinental Triangulation and the American Arc of the Parallel.

nearly correct to say it is an oblate spheroid. This, however, is not strictly accurate, as is shown in the succeeding discussion.

THE EARTH A GEOID

Conditions Producing Irregularities. If the earth had been made up of the same kinds of material uniformly distributed throughout its mass, it would probably have assumed, because of its rotation, the form of a regular oblate spheroid. But the earth has various materials unevenly distributed in it, and this has led to many slight variations from regularity in form.

Equator Elliptical. Pendulum experiments and measurements indicate not only that meridians are elliptical but that the equator itself may be slightly elliptical, its longest axis passing through the earth from 15° E. to 165° W. and its shortest axis from 105° E. to 75° W. The amount of this oblateness of the equator is estimated at about $\frac{1}{4,000}$ or a difference of two miles in the lengths of these two diameters of the equator.

Thus the meridian circle passing through central Africa and central Europe (15° E.) and around near Behring Strait (165° W.) may be slightly more oblate than the other meridian circles, the one which is most nearly circular passing through central Asia (105° E.), eastern North America, and western South America (75° W.).

United States Curved Unequally. It is interesting to note that the dimensions of the degrees of the long arc of the 39th parallel surveyed in the United States bear out to a remarkable extent the theory that the earth is slightly attened longitudinally, making it even more than that just given, which was calculated by Sir John Herschel and A. R. Clarke. The average length of degrees of longitude from the

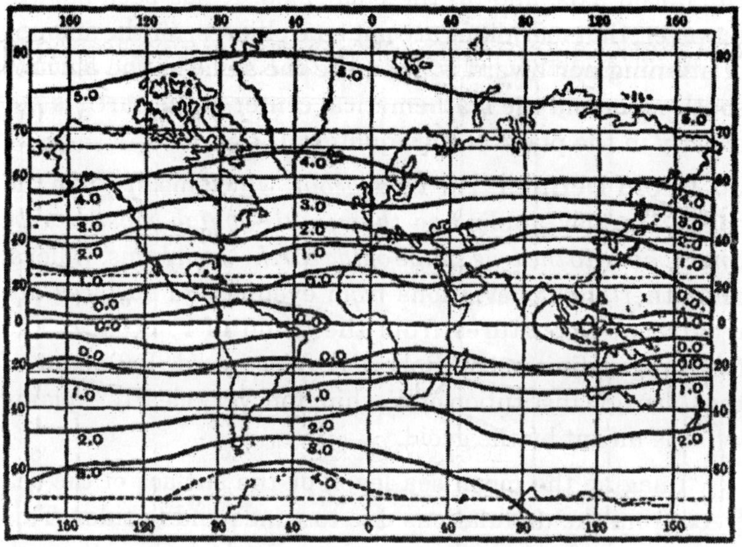

Fig. 13. Gravimetric lines showing variation in force of gravity

Atlantic coast for the first 1, 500 miles corresponds closely to the Clarke table, and thus those degrees are longer, and the rest of the arc corresponds closely to the Bessel table and shows shorter degrees.

	Diff. in long.	Length of 1°	Clarke	Bessel
Cape May to Wallace (Kansas)	26.661°	53.829 mi.	53.828 mi	....
Wallace to Uriah (Calif.)..........	21.618°	53.822 mi	....	53.821 mi.

Earth not an Ellipsoid of Three Unequal Axes. This oblateness of the meridians and oblateness of the equator led some to treat the earth as an ellipsoid of three unequal axes: (1) the longest equatorial axis, (2) the shortest equatorial axis, and (3) the polar axis. It has been shown, however, that meridians are not true ellipses, for the amount of attening northward is not quite the same as the amount southward, and the mathematical center of the earth is not exactly in the plane of the equator.

Geoid Defined. The term *geoid*, which means "like the earth," is now applied to *that figure which most nearly corresponds to the true shape of the earth*. Mountains, valleys, and other slight deviations from evenness of surfaces are treated as departures from the geoid of reference. The following definition by Robert S. Woodward, President of the Carnegie Institution of Washington, very clearly explains what is meant by the geoid.*

"Imagine the mean sea level, or the surface of the sea freed from the undulations due to winds and to tides. This mean sea surface, which may be conceived to extend through the continents, is called the geoid. It does not coincide exactly with the earth's spheroid, but is a slightly wavy surface lying partly above and partly below the spheroidal surface, by small but as yet not definitely known amounts. The determination of the geoid is now one of the most important problems of geophysics."

An investigation is now in progress in the United States for determining a new geoid of reference upon a plan never followed hitherto. The following is a lucid description† of the plan by John F. Hayford, Inspector of Geodetic Work, United States Coast and Geodetic Survey.

Area Method of Determining Form of the Earth. "The arc method of deducing the figure of the earth may be illustrated by supposing that a skilled workman to whom is given several stiff wires, each representing a geodetic arc, either of a parallel or a meridian, each bent to the radius

*Encyclopaedia Americana.

† Given at the International Geographic Congress, 1904.

deduced from the astronomic observations of that arc, is told in what latitude each is located on the geoid and then requested to construct the ellipsoid of revolution which will conform most closely to the bent wires. Similarly, the area method is illustrated by supposing that the workman is given a piece of sheet metal cut to the outline of the continuous triangulation which is supplied with necessary astronomic observations, and accurately molded to fix the curvature of the geoid, as shown by the astronomic observations, and that the workman is then requested to construct the ellipsoid of revolution which will conform most accurately to the bent sheet. Such a bent sheet essentially includes within itself the bent wires referred to in the first illustration, and, moreover, the wires are now held rigidly in their proper relative positions. The sheet is much more, however, than this rigid system of bent lines, for each arc usually treated as a line is really a belt of considerable width which is now utilized fully. It is obvious that the workman would succeed much better in constructing accurately the required ellipsoid of revolution from the one bent sheet than from the several bent wires. When this proposition is examined analytically it will be seen to be true to a much greater extent than appears from this crude illustration."

"The area of irregular shape which is being treated as a single unit extends from Maine to California and from Lake Superior to the Gulf of Mexico. It covers a range of 57° in longitude and 19° in latitude, and contains 477 astronomic stations. This triangulation with its numerous accompanying astronomical observations will, even without combination with similar work in other countries, furnish a remarkably strong determination of the figure and size of the earth."

It is possible that at some distant time in the future the dimensions and form of the geoid will be so accurately known that instead of using an oblate spheroid of reference (that is, a spheroid of such dimensions as most closely correspond to the earth, treated as an oblate spheroid such as the Clarke Spheroid of 1866), as is now done, it will be possible to treat any particular area of the earth as having its own peculiar curvature and dimensions.

Conclusion. What is the form of the earth? We went to considerable pains to prove that the earth is a sphere. That may be said to be its general form, and in very many calculations it is always so treated. For more exact calculations, the earth's departures from a sphere must be borne in mind. The regular geometric solid which the earth most nearly resembles is an oblate spheroid. Strictly speaking, however, the form of the earth (not considering such irregularities as mountains and valleys) must be called a *geoid*.

DIRECTIONS ON THE EARTH

On a Meridian Circle. Think of yourself as standing on a great circle of the earth passing through the poles. Pointing from the northern horizon by way of your feet to the southern horizon, you have pointed to all parts of the meridian circle beneath you. Your arm has swung through an angle of 180° but you have pointed through all points of the meridian circle, or 360° of it. Drop your arm 90° or from the horizon to the nadir, and you have pointed through half of the meridian circle, for 180° of latitude. It is apparent, then, that for every degree you drop your arm, you point through a space of two degrees of latitude upon the earth beneath.

The north pole is, let us say, 45° from you. Drop your arm 22½° from the northern horizon, and you will point directly toward the north pole (Fig. 14). *Whatever your latitude, drop your arm half as many degrees from the* *northern horizon as you are degrees from the pole, and you will point directly toward that pole.**

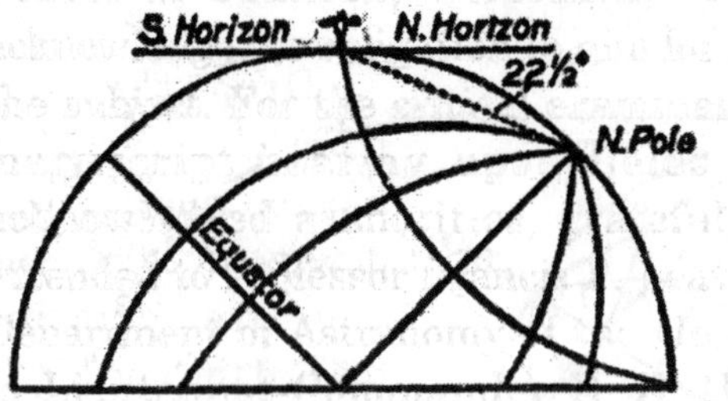

Fig. 14

You may be so accustomed to thinking of the north pole as northward in a horizontal line from you that it does not seem real to think of it as below the horizon. This is because one is liable to forget that he is living on a ball. To point to the horizon is to point away from the earth.

A Pointing Exercise. It may not be easy or even essential to learn exactly to locate many places in relation to the home region, but the ability to locate readily some salient points greatly clarifies one's sense of location and conception of the earth as a ball.

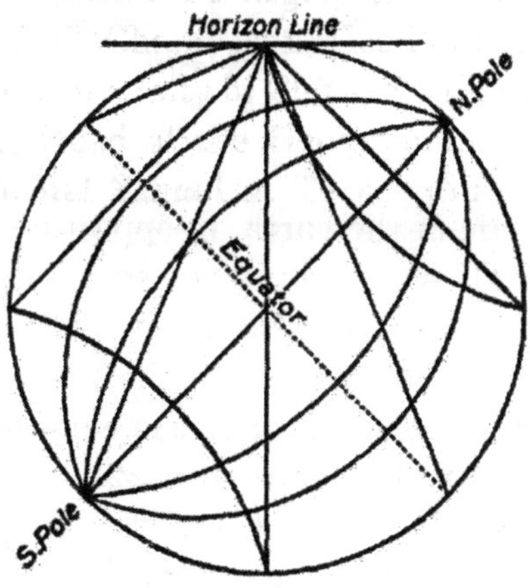

Fig. 15

The following exercise is designed for students living not far from the 45th parallel. Since it is impossible to point the arm or pencil with accuracy at any given angle, it is roughly adapted for the north temperate latitudes (Fig. 15). Persons living in

*The angle included between a tangent and a chord is measured by one half the intercepted arc.

the southern states may use Figure 16, based on the 30th parallel. The student should make the necessary readjustment for his own latitude.

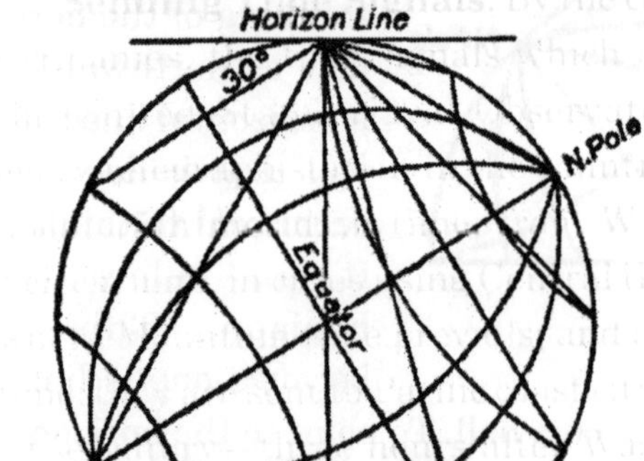

Fig. 16

Drop the arm from the northern horizon quarter way down, or 22½°, and you are pointing toward the north pole (Fig. 15). Drop it half way down, or 45° from the horizon, and you are pointing 45° the other side of the north pole, or half way to the equator, on the same parallel but on the opposite side of the earth, in opposite longitude. Were you to travel half way around the earth in a due easterly or westerly direction, you would be at that point. Drop the arm 22½° more, or 67½° from the horizon, and you are pointing 45° farther south or to the equator on the opposite side of the earth. Drop the arm 22½° more, or 90° from the horizon, toward your feet, and you are pointing toward our antipodes, 45° south of the equator on the meridian opposite ours. Find where on the earth this point is. Is the familiar statement, "digging through the earth to China," based upon a correct idea of positions and directions on the earth?

From the southern horizon drop the arm 22½°, and you are pointing to a place having the same longitude but on the equator. Drop the arm 22½° more, and you point to a place having the same longitude as ours but opposite latitude, being 45° south of the equator on our meridian. Drop the arm 22½° more, and you point toward the south pole. Practice until you can point directly toward any of these seven points without reference to the diagram.

LATITUDE AND LONGITUDE

Origin of Terms. Students often have difficulty in remembering whether it is latitude that is measured east and west, or longitude. When we recall the fact that to the people who first used these terms the earth was believed to be longer east and west than north and south, and now we know that owing to the oblateness of the earth this is actually the case, we can easily remember that longitude (from the Latin *longus*, long) is measured east and west. The word latitude is from the Latin *latitudo*, which is from *latus*, wide, and was originally used to designate measurement of the "width of the earth," or north and south.

Antipodal Areas. From a globe one can readily ascertain the point which is exactly opposite any given one on the earth. The map showing antipodal areas indicates at a glance what portions of the earth are opposite each other; thus Australia lies directly through the earth from mid-Atlantic, the point antipodal to Cape Horn is in central Asia, etc.

Longitude is measured on parallels and is reckoned from some meridian selected as standard, called the *prime meridian*. The meridian which passes through the Royal Observatory at Greenwich, near London, has long been the prime meridian most used. In many countries the meridian passing through the capital is taken as the prime meridian. Thus, the Portuguese use the meridian of the Naval Observatory in the Royal Park at Lisbon, the French that of the Paris Observatory, the Greeks that of the Athens

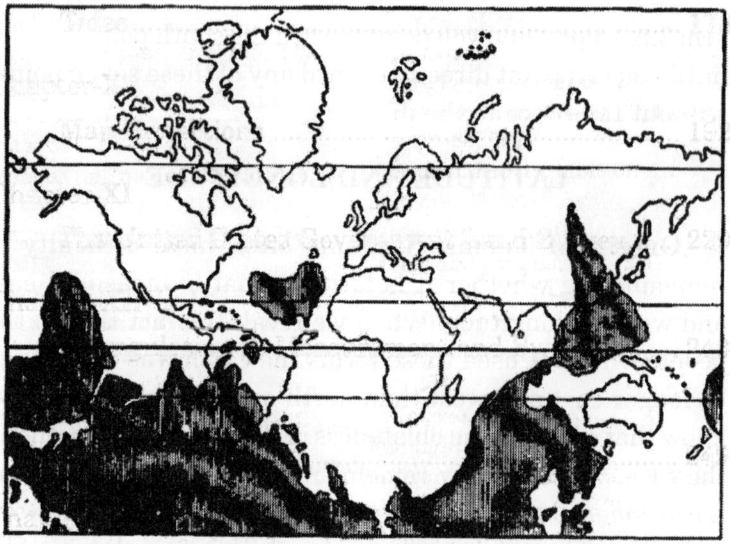

Fig. 17. Map of Antipodal Areas

Observatory, the Russians that of the Royal Observatory at Pulkowa, near St. Petersburg.

In the maps of the United States the longitude is often reckoned both from Greenwich and Washington. The latter city being a trifle more than 77° west of Greenwich, a meridian numbered at the top of the map as 90° west from Greenwich, is numbered at the bottom as 13° west from Washington. Since the United States Naval Observatory, the point in Washington reckoned from, is 77° 3' 81'' west from Greenwich, this is slightly inaccurate. Among all English speaking people and in most nations of the world, unless otherwise designated, the longitude of a place is understood to be reckoned from Greenwich.

The longitude of a place is the arc of the parallel intercepted between it and the prime meridian. Longitude may also be defined as the arc of the equator intercepted between the prime meridian and the meridian of the place whose longitude is sought.

Since longitude is measured on parallels, and parallels grow smaller toward the poles, degrees of longitude are shorter toward the poles, being degrees of smaller circles.

Latitude is measured on a meridian and is reckoned from the equator. The number of degrees in the arc of a meridian circle, from the place whose latitude is sought to the equator, is its latitude. Stated more formally, the latitude of a place is the arc of the meridian intercepted between the equator and that place. (See Latitude in Glossary.) What is the greatest number of degrees of latitude any place may have? What places have no latitude?

Comparative Lengths of Degrees of Latitude. If the earth were a perfect sphere, meridian circles would be true mathematical circles. Since the earth is an oblate spheroid, meridian circles, so called, curve less rapidly toward the poles. Since the curvature is greatest near the equator, one would have to travel less distance on a meridian there to cover a degree of curvature, and a degree of latitude is thus shorter

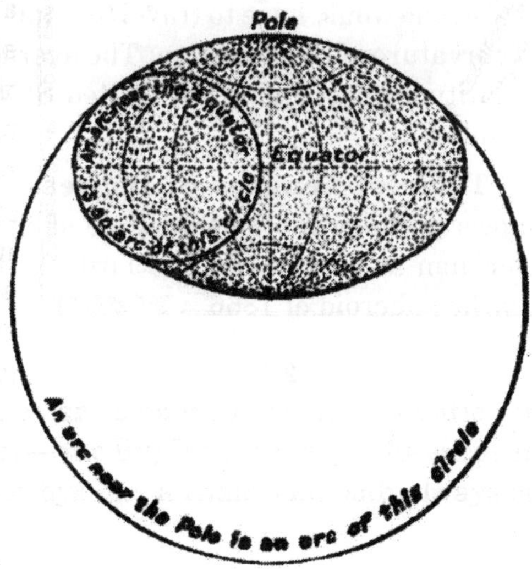

Fig. 18

near the equator. Conversely, the meridian being slightly attened toward the poles, one would travel farther there to cover a degree of latitude, hence degrees of latitude are longer toward the poles. Perhaps this may be seen more clearly from Figure 18.

While all circles have 360°, the degrees of a small circle are, of course, shorter than the degrees of a greater circle. Now an arc of a meridian near the equator is obviously a part of a smaller circle than an arc taken near the poles and, consequently, the degrees are shorter. Near the poles, because of the flatness of a meridian there, an arc of a meridian is a part of a larger circle and the degrees are longer. As we travel northward, the North star (polestar) rises from the horizon. In traveling from the equator on a meridian, one would go 68.7 miles to see the polestar rise one degree, or, in other words, to cover one degree of curvature of the meridian. Near the pole, where the earth is attest, one would have to travel 69.4 miles to cover one degree of curvature of the meridian. The average length of a degree of latitude throughout the United States is almost exactly 69 miles.

Table of Lengths of Degrees. The following table shows the length of each degree of the parallel and of the meridian at every degree of latitude. It is based upon the Clarke spheroid of 1866.

Lat.	Deg. Par. Miles	Deg. Mer. Miles	Deg. Lat.	Deg. Par. Miles	Deg. Mer. Miles	Deg. Lat.	Par. Miles	Mer. Miles
0°	69:172	68:704	31.	59:345	68:890	61.	33:623	69:241
1	69:162	68:704	32	58:716	68:901	62	32:560	69:251
2	69:130	68:705	33	58:071	68:912	63	31:488	69:261
3	69:078	68:706	34	57:407	68:923	64	30:406	69:271
4	69:005	68:708	35	56:725	68:935	65	29:315	69:281
5	68:911	68:710	36	56:027	68:946	66	28:215	69:290
6	68:795	68:712	37	55:311	68:958	67	27:106	69:299
7	68:660	68:715	38	54:579	68:969	68	25:988	69:308
8	68:504	68:718	39	53:829	68:981	69	24:862	69:316
9	68:326	68:721	40	53:063	68:993	70	23:729	69:324
10	68:129	68:725	41	52:281	69:006	71	22:589	69:332
11	67:910	68:730	42	51:483	69:018	72	21:441	69:340
12	67:670	68:734	43	50:669	69:030	73	20:287	69:347
13	67:410	68:739	44	49:840	69:042	74	19:127	69:354
14	67:131	68:744	45	48:995	69:054	75	17:960	69:360
15	66:830	68:751	46	48:136	69:066	76	16:788	69:366
16	66:510	68:757	47	47:261	69:079	77	15:611	69:372
17	66:169	68:764	48	46:372	69:091	78	14:428	69:377
18	65:808	68:771	49	45:469	69:103	79	13:242	69:382
19	65:427	68:778	50	44:552	69:115	80	12:051	69:386
20	65:026	68:786	51	43:621	69:127	81	10:857	69:390
21	64:606	68:794	52	42:676	69:139	82	9:659	69:394
22	64:166	68:802	53	41:719	69:151	83	8:458	69:397
23	63:706	68:811	54	40:749	69:163	84	7:255	69:400
24	63:228	68:820	55	39:766	69:175	85	6:049	69:402
25	62:729	68:829	56	38:771	69:186	86	4:842	69:404
26	62:212	68:839	57	37:764	69:197	87	3:632	69:405
27	61:676	68:848	58	36:745	69:209	88	2:422	69:407
28	61:122	68:858	59	35:716	69:220	89	1:211	69:407
29	60:548	68:869	60	34:674	69:230	90	0:000	69:407
30	59:956	68:879						

CHAPTER-III

THE ROTATION OF THE EARTH

THE CELESTIAL SPHERE

Apparent Dome of the Sky. On a clear night the stars twinkling all over the sky seem to be fixed in a dark dome fitting down around the horizon. This apparent concavity, studded with heavenly bodies, is called the celestial sphere. Where the horizon is free from obstructions, one can see half* of the celestial sphere at a given time from the same place.

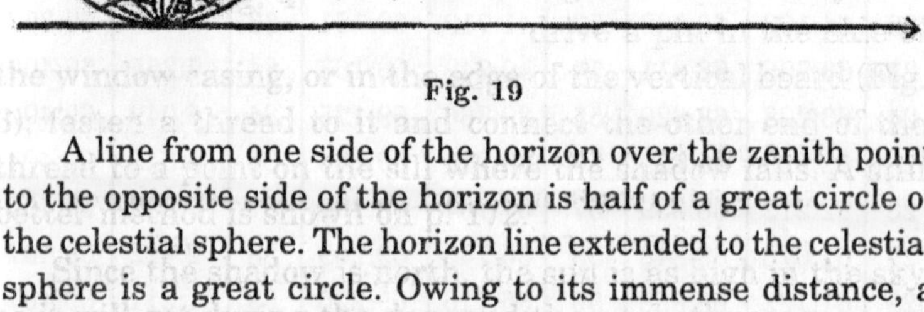

Fig. 19

A line from one side of the horizon over the zenith point to the opposite side of the horizon is half of a great circle of the celestial sphere. The horizon line extended to the celestial sphere is a great circle. Owing to its immense distance, a

*No allowance is here made for the refraction of rays of light or the slight curvature of the globe in the locality.

line from an observer at A (Fig. 19), pointing to a star, will make a line apparently parallel to one from *B* to the same star. The most refined measurements at present possible fail to show any angle whatever between them.

We may note the following in reference to the celestial sphere.

(1) The earth seems to be a mere point in the center of this immense hollow sphere. (2) The stars, however distant, are apparently fixed in this sphere. (3) Any plane from the observer, if extended, will divide the celestial sphere into two equal parts. (4) Circles may be projected on this sphere and positions on it indicated by degrees in distance from established circles or points.

Celestial Sphere seems to Rotate. The earth rotates on its axis (the term rotation applied to the earth refers to its daily or axial motion). To us, however, the earth seems stationary and the celestial sphere seems to rotate. Standing in the center of a room and turning one's body around, the objects in the room seem to rotate around in the opposite direction. The point overhead will be the only one that is stationary. Imagine a fly on a rotating sphere. If it were on one of the poles, that is, at the end of the axis of rotation, the object directly above it would constantly remain above it while every other fixed object would seem to swing around in circles. Were the fly to walk to the equator, the point directly away from the globe would cut the largest circle around him and the stationary points would be along the horizon.

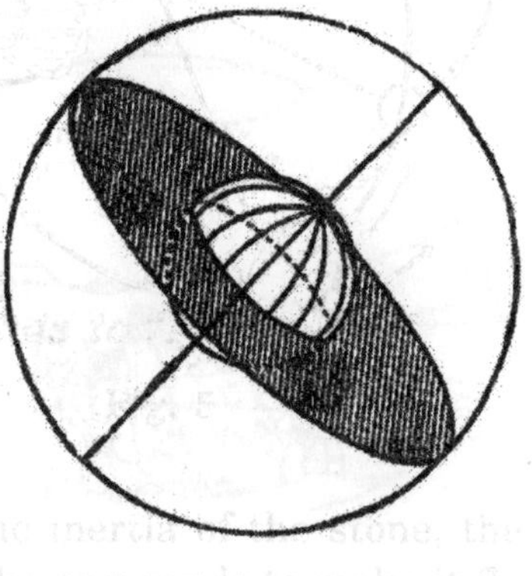

Fig. 20

Celestial Pole. The point in the celestial sphere directly above the pole and in line with the axis has no motion. It is called the celestial pole. The star nearest the pole of the celestial sphere and directly above the north pole of the earth is called the North star, and the star nearest the southern celestial pole the South star. It may be of interest to note that as we located the North star by reference to the Big Dipper, the South star is located by reference to a group of stars known as the Southern Cross.

Celestial Equator. A great circle is conceived to extend around the celestial sphere 90° from the poles (Fig. 20). This is called the *celestial equator*. The axis of the earth, if prolonged, would pierce the celestial poles, almost pierce the North and South stars, and the equator of the earth if extended would coincide with the celestial equator.

Fig. 21

At the North Pole. An observer at the north pole will see the North star almost exactly overhead, and as the earth turns around under his feet it will remain constantly overhead (Fig. 21). Half way, or 90° from the North star, is the celestial equator around the horizon. As the earth rotates,—though it seems to us perfectly still,—the stars around the sky seem to swing in circles in the opposite direction. The planes of the star paths are parallel to the horizon. The same half of the celestial sphere can be seen all of the time, and stars below the horizon always remain so.

All stars south of the celestial equator being forever invisible at the north pole, Sirius, the brightest of the stars, and many of the beautiful constellations, can never be seen from that place. How peculiar the view of the heavens must be from the pole, the Big Dipper, the Pleiades, the Square of Pegasus, and other star groups swinging eternally around in courses parallel to the horizon. When the sun, moon, and planets are in the portion of their courses north of the celestial equator, they, of course, will be seen throughout continued rotations of the earth until they pass below the celestial equator, when they will remain invisible again for long periods.

The direction of the daily apparent rotation of the stars is from left to right (westward), the direction of the hands of a clock looked at from above. Lest the direction of rotation at the North pole be a matter of memory rather than of insight, we may notice that in the United States and Canada when we face southward we see the sun's daily course in the direction left to right (westward), and going poleward the direction remains the same though the sun approaches the horizon more and more as we approach the North pole.

At the South Pole. An observer at the South pole, at the other end of the axis, will see the South star directly overhead, the celestial equator on the horizon, and the plane of the star circles parallel with the horizon. The direction of the apparent rotation of the celestial sphere is

Fig. 22

from right to left, counter-clockwise. If a star is seen at one's right on the horizon at six o'clock in the morning, at noon it will be in front, at about six o'clock at night at his left, at midnight behind him, and at about six o'clock in the morning at his right again.

At the Equator. An observer at the equator sees the stars in the celestial sphere to be very different in their positions in relation to himself. Remembering that he is standing with the line of his body at right angles to the axis of the earth, it is easy to understand why all the stars of the celestial sphere seem to be shifted around 90° from where they were at the poles. The celestial equator is a great circle extending from east to west directly overhead. The North star is seen on the northern horizon and the South star on the southern horizon. The planes of the circles followed by stars in their daily orbits cut the horizon at right angles, the horizon being parallel to the axis. At the equator one can see the entire celestial sphere, half at one time and the other half about twelve hours later.

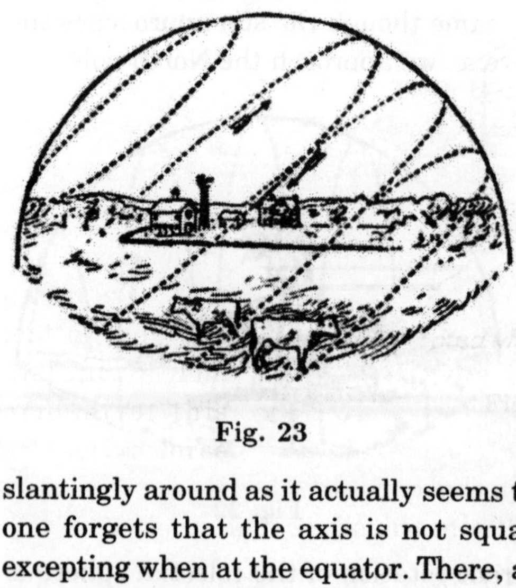

Fig. 23

Between Equator and Poles. At places between the equator and the poles, the observer is liable to feel that a star rising due east ought to pass the zenith about six hours later instead of swinging slantingly around as it actually seems to do. This is because one forgets that the axis is not squarely under his feet excepting when at the equator. There, and there only, is the axis at right angles to the line of one's body when erect. The apparent rotation of the celestial sphere is at right angles to the axis.

Photographing the Celestial Sphere. Because of the earth's rotation, the entire celestial sphere seems to rotate. Thus we see stars daily circling around, the polestar always stationary. When stars are photographed, long exposures are necessary that their faint light may affect the sensitive plate of the camera, and the photographic instruments must be constructed so that they will move at the same rate and in the same direction as the stars, otherwise the stars will leave trails on the plate. When the photographic instrument thus follows the stars in their courses, each is shown as a speck on the plate and comets, meteors, planets, or asteroids, moving at different rates and in different directions, show as traces.

Rotation of Celestial Sphere is Only Apparent. For a long time it was believed that the heavenly bodies rotated around the stationary earth as the center. It was only about five hundred years ago that the astronomer Copernicus established the fact that the motion of the sun and stars around the earth is only apparent, the earth rotating. We may be interested in some proofs that this is the case. It seems hard to believe at first that this big earth, 25,000 miles in circumference can turn around once in a day. "Why, that would give us a whirling motion of over a thousand miles an hour at the equator." "Who could stick to a merry-go-round going at the rate of a thousand miles an hour?" When we see, however, that the sun 93,000,000 miles away, would have to swing around in a course of over 580,000,000 miles per day, and the stars, at their tremendous distances, would have to move at unthinkable rates of speed, we see that it is far easier to believe that it is the earth and not the celestial sphere that rotates daily. We know by direct observation that other planets, the sun and the moon, rotate upon their axis, and may reasonably infer that the earth does too.

So far as the whirling motion at the equator is concerned, it does give bodies a slight tendency to fly off, but the amount of this force is only $\frac{1}{289}$ as great as the attractive influence of the earth; that is, an object which would weigh 289 pounds at the equator, were the earth at rest, weighs a pound less because of the centrifugal force of rotation (see p. 14).

PROOFS OF THE EARTH'S ROTATION

Eastward Deflection of Falling Bodies. Perhaps the simplest proof of the rotation of the earth is one pointed out by Newton, although he had no means of demonstrating it. With his clear vision he said that if the earth rotates and an object were dropped from a considerable height, instead of falling directly toward the center of the earth in the direction of the plumb line,* it would be deflected toward the east. Experiments have been made in the shafts of mines where air currents have been shut off and a slight but unmistakable eastward tendency has been observed.

During the summer of 1906, a number of newspapers and magazines in the United States gave accounts of the eastward falling of objects dropped in the deep mines of northern Michigan, one of which (Shaft No. 3 of the Tamarack mine) is the deepest in the world, having a vertical depth of over one mile (and still digging!). It was stated that objects dropped into such a shaft never reached the bottom but always lodged among timbers on the east side. Some papers added a touch of the grewsome by implying that among the objects found clinging to the east side are "pieces

*The slight geocentric deviations of the plumb line are explained on p. 280.

of a dismembered human body" which were not permitted to fall to the bottom because of the rotation of the earth. Following is a portion of an account* by F. W. McNair, President of the Michigan College of Mines.

McNair's Experiment. "Objects dropping into the shaft under ordinary conditions nearly always start with some horizontal velocity, indeed it is usually due to such initial velocity in the horizontal that they get into the shaft at all. Almost all common objects are irregular in shape, and, drop one of them ever so carefully, contact with the air through which it is passing soon deviates it from the vertical, giving it a horizontal velocity, and this when the air is quite still. The object slides one way or another on the air it compresses in front of it. Even if the body is a sphere, the air will cause it to deviate, if it is rotating about an axis out of the vertical. Again, the air in the shaft is in ceaseless motion, and any obliquity of the currents would obviously deviate the falling body from the vertical, no matter what its shape. If the falling object is of steel, the magnetic influence of the air mains and steam mains which pass down the shaft, and which invariably become strongly magnetic, may cause it to swerve from a vertical course...

"A steel sphere, chosen because it was the only convenient object at hand, was suspended about one foot from the timbers near the western corner of the compartment. The compartment stands diagonally with reference to the cardinal points. Forty-two hundred feet below a clay bed was placed, having its eastern edge some five feet east of the point of suspension of the ball. When the ball appeared to be still the suspending thread was burned, and the instant of the dropping of the ball was indicated by a prearranged signal transmitted by telephone to the observers below, who,

*In the Mining and Scientific Press, July 14, 1906.

watch in hand, waited for the sphere to strike the bed of clay. It failed to appear at all. Another like sphere was hung in the center of the compartment and the trial was repeated with the same result. The shaft had to be cleared and no more trials were feasible. Some months later, one of the spheres, presumably the latter one, was found by a timberman where it had lodged in the timbers 800 feet from the surface.

Fig. 24

"It is not probable, however, that these balls lodged because of the earth's rotation alone... The matter is really more complicated than the foregoing discussion implies. It has received mathematical treatment from the great Gauss. According to his results, the deviation to the east for a fall of 5,000 feet at the Tamarack mine should be a little under three feet. Both spheres had that much to spare before striking the timbers. It is almost certain, therefore, that others of the causes mentioned in the beginning acted to prevent a vertical fall. At any rate, these trials serve to emphasize the unlikelihood that an object which falls into a deep vertical shaft, like those at the Tamarack mine, will reach the bottom, even when some care is taken in selecting it and also to start it vertically.

"If the timbering permits lodgment, as is the case in most shafts, it may truthfully be said that if a shaft is deep in proportion to its cross section few indeed will be the objects falling into it which will reach the bottom, and such objects are more likely to lodge on the easterly side than on any other."

The Foucault Experiment. Another simple demonstration of the earth's rotation is by the celebrated Foucault experiment. In 1851, the French physicist, M. Leon Foucault, suspended from the dome of the Pantheon, in Paris, a heavy iron ball by wire two hundred feet long. A pin was fastened to the lowest side of the ball so that when swinging it traced a slight mark in a layer of sand placed beneath it. Carefully the long pendulum was set swinging. It was found that the path gradually moved around toward the right. Now either the pendulum changed its plane or the building was gradually turned around. By experimenting with a ball suspended from a ruler one can readily see that gradually turning the ruler will not change the plane of the swinging pendulum. If the pendulum swings back and forth in a north and south direction, the ruler can be entirely turned around without changing the direction of the pendulum's swing. If at the north pole a

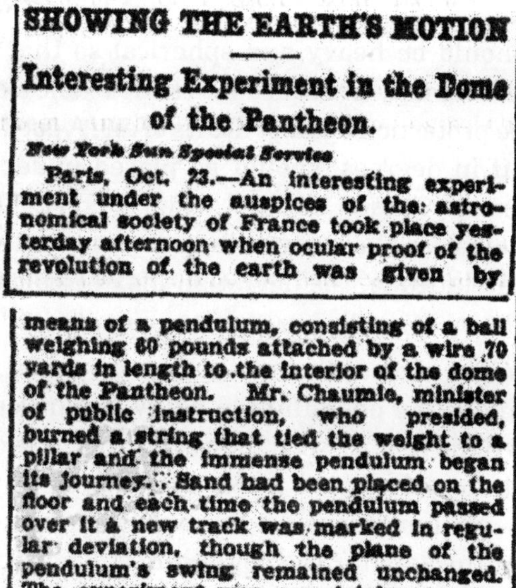

SHOWING THE EARTH'S MOTION

Interesting Experiment in the Dome of the Pantheon.

New York Sun Special Service

Paris, Oct. 23.—An interesting experiment under the auspices of the astronomical society of France took place yesterday afternoon when ocular proof of the revolution of the earth was given by means of a pendulum, consisting of a ball weighing 60 pounds attached by a wire 70 yards in length to the interior of the dome of the Pantheon. Mr. Chaumie, minister of public instruction, who presided, burned a string that tied the weight to a pillar and the immense pendulum began its journey. Sand had been placed on the floor and each time the pendulum passed over it a new track was marked in regular deviation, though the plane of the pendulum's swing remained unchanged. The experiment was completely successful.

–1902

Fig. 25

pendulum was set swinging toward a fixed star, say Arcturus, it would continue swinging toward the same star and the earth would thus be seen to turn around in a day. The earth would not seem to turn but the pendulum would seem to deviate toward the right or clockwise.

Conditions for Success. The Foucault experiment has been made in many places at different times. To be successful there should be a long slender wire, say forty feet or more in length, down the well of a stairway. The weight suspended should be heavy and spherical so that the impact against the air may not cause it to slide to one side, and there should be protection against drafts of air. A good sized circle, marked off in degrees, should be placed under it, with the center exactly under the ball when at rest. From the rate of the deviation the latitude may be easily determined or, knowing the latitude, the deviation may be calculated.

To Calculate Amount of Deviation. At first thought it might seem as though the floor would turn completely around under the pendulum in a day, regardless of the latitude. It will be readily seen, however, that it is only at the pole that the earth would make one complete rotation under the pendulum in one day* or show a deviation of 15° in an hour. At the equator the pendulum will show no deviation, and at intermediate latitudes the rate of deviation varies. Now the ratio of variation from the pole considered as *one* and the equator as *zero* is shown in the table of "natural sines" (p. 310). It can be demonstrated that the number of degrees the plane of the pendulum will deviate in one hour at any latitude is found by multiplying 15° by the sine of the latitude.

*Strictly speaking, in one sidereal day.

d = deviation

ϕ = latitude

$\therefore \quad d = \sin \phi \times 15°$

Whether or not the student has a very clear conception of what is meant by "the sine of the latitude" he may easily calculate the deviation or the latitude where such a pendulum experiment is made.

Example. Suppose the latitude is 40° Sine 40° = .6428. The hourly deviation at that latitude, then, is .6428 × 15° or 9.64° Since the pendulum deviates 9.64° in one hour, for the entire circuit it will take as many hours as that number of degrees is contained in 360° or about $37\frac{1}{3}$ hours. It is just as simple tocalculate one's latitude if the amount of deviation for one hour is known. Suppose the plane of the pendulum is observed to deviate 9° in an hour.

Sine of the latitude × 15° = 9°.

$\therefore$ Sine of the latitude = $\frac{9}{15}$ or .6000.

From the table of sines we find that this sine, .6000, corresponds more nearly to that of 37° (.6018) than to the sine of any other whole degree, and hence 37° is the latitude where the hourly deviation is 9°. At that latitude it would take forty hours (360 ÷ 9 = 40) for the pendulum to make the entire circuit.

Table of Variations. The following table shows the deviation of the plane of the pendulum for one hour and the time required to make one entire rotation.

Latitude.	Hourly Deviation.	Circuit of Pendulum.	Latitude.	Hourly Deviation.	Circuit of Pendulum.
5°	1.31°	275 hrs.	50°	11.49°	31 hrs.
10	2.60	138	55	12.28	29
15	3.88	93	60	12.99	28
20	5.13	70	65	13.59	26
25	6.34	57	70	14.09	25.5
30	7.50	48	75	14.48	24.8
35	8.60	42	80	14.77	24.5
40	9.64	37	85	14.94	24.1
45	10.61	34	90	15.00	24.0

Other Evidence. Other positive evidence of the rotation of the earth we have in the fact that the equatorial winds north of the equator veer toward the east and polar winds toward the west—south of the equator exactly opposite—and this is precisely the result which would follow from the earth's rotation. Cyclonic winds in the northern hemisphere in going toward the area of low pressure veer toward the right and anti-cyclonic winds also veer toward the right in leaving areas of high pressure, and in the southern hemisphere their rotation is the opposite. No explanation of these well-known facts has been satisfactorily advanced other than the eastward rotation of the earth, which easily accounts for them.

Perhaps the best of modern proofs of the rotation of the earth is demonstrated by means of the spectroscope. A discussion of this is reserved until the principles are explained (p. 108) in connection with the proofs of the earth's revolution.

VELOCITY OF ROTATION

The velocity of the rotation at the surface, in miles per hour, in different latitudes, is as follows:

Latitude.	Velocity.	Latitude.	Velocity.	Latitude.	Velocity.
0	1038	44	748	64	456
5	1034	45	735	66	423
10	1022	46	722	68	390
15	1002	47	709	70	356
20	975	48	696	72	322
25	941	49	682	74	287
30	899	50	668	76	252
32	881	51	654	78	216
34	861	52	640	80	181
36	840	53	626	82	145
38	819	54	611	84	109
39	807	55	596	86	73
40	796	56	582	88	36
41	784	58	551	89	18
42	772	60	520	89½	9
43	760	62	488	90	0

Uniform Rate of Rotation. There are theoretical grounds for believing that the rate of the earth's rotation is getting gradually slower. As yet, however, not the slightest variation has been discovered. Before attacking the somewhat complex problem of time, the student should clearly bear in mind the fact that the earth rotates on its axis with such unerring regularity that this is the most perfect standard for any time calculations known to science.

DETERMINATION OF LATITUDE

Altitude of Celestial Pole Equals Latitude. Let us return, in imagination, to the equator. Here we may see the North star on the horizon due north of us, the South star on the horizon due south, and halfway between these two points, extending from due east through the zenith to due west, is the celestial equator. If we travel northward we shall be able to see objects which were heretofore hidden from view by the curvature of the earth. We shall find that the South star becomes hidden from sight for the same reason and the North star seems to rise in the sky. The celestial equator no longer extends through the point directly overhead but is somewhat to the south of the zenith, though it still intersects the horizon at the east and west points. As we go farther north this rising of the northern sky and sinking of the southern sky becomes greater. If we go halfway to the north pole we shall find the North star halfway between the zenith and the northern horizon, or at an altitude of 45° above the horizon. For every degree of curvature of the earth we pass over, going northward, the North star rises one degree from the horizon. At New Orleans the North star is 30° from the horizon, for the city is 30° from the equator. At Philadelphia, 40. north latitude, the North star is 40° from the horizon. South of the equator the converse of this is true. The North star sinks from the horizon and the South star rises as one travels southward from the equator. *The altitude of the North star is the latitude of a place north of the equator and the altitude of the South star is the latitude of a place south of the equator.* It is obvious, then, that the problem of determining latitude is the problem of determining the altitude of the celestial pole.

To Find Your Latitude. By means of the compasses and scale, ascertain the altitude of the North star. This can be done by placing one side of the compasses on a level window sill and sighting the other side toward the North star, then measuring the angle thus formed. Another simple process for ascertaining latitude is to determine the altitude of a star not far from the North star when it is highest and when it is lowest; the average of these altitudes is the altitude of the pole, or the latitude. This may easily be done in latitudes north of 38° during the winter, observing, say, at 6 o'clock in the morning and at 6 o'clock in the evening. This is simple in that it requires no tables. Of course such measurements are very crude with simple instruments, but with a little care one will usually be surprised at the accuracy of his results.

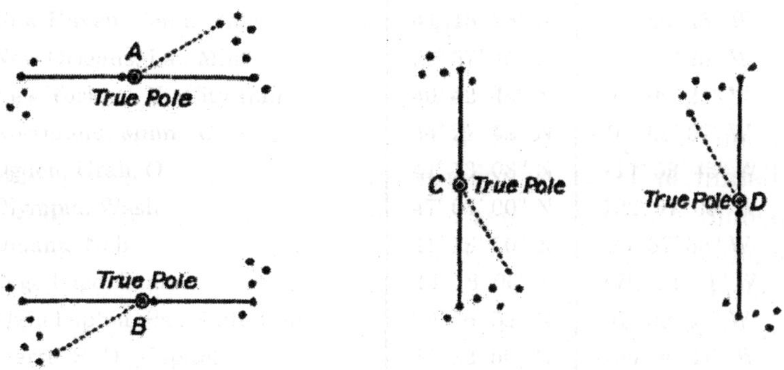

Fig. 26

Owing to the fact that the North star is not exactly at the north pole of the celestial sphere, it has a slight rotary motion. It will be more accurate, therefore, if the observation is made when the Big Dipper and Cassiopeia are in one of the positions (*A* or *B*) represented by Figure 26. In case of these positions the altitude of the North star will give the true latitude, it then being the same altitude as the pole of the celestial sphere. In case of position *D*, the North star is about 1¼° below the true pole, hence 1¼° must be added to the altitude of the star. In case of position *C*, the North star is 1¼° above the true pole, and that amount must be subtracted from its altitude. It is obvious from the diagrams that a true north and south line can be struck when the stars are in positions *C* and *D*, by hanging two plumb lines so that they lie in the same plane as the zenith meridian line through Mizar and Delta Cassiopeiæ. Methods of determining latitude will be further discussed on pp. 172–175. The instrument commonly used in observations for determining latitude is the meridian circle, or, on shipboard, the sextant. Read the description of these instruments in any text on astronomy.

QUERIES

In looking at the heavenly bodies at night do the stars, moon, and planets all look as though they were equally distant, or do some appear nearer than others? The fact that people of ancient times believed the celestial sphere to be made of metal and all the heavenly bodies fixed or moving therein, would indicate that to the observer who is not biased by preconceptions, all seem equally distant. If they did not seem equally distant they would not assume the apparently spherical arrangement.

The declination, or distance from the celestial equator, of the star (Benetnasch) at the end of the handle of the Big Dipper is 50°. How far is it from the celestial pole? At what latitude will it touch the horizon in its swing under the North star? How far south of the equator could one travel and still see that star at some time?

CHAPTER-IV

LONGITUDE AND TIME

SOLAR TIME

Sun Time Varies. The sun is the world's great timekeeper. He is, however, a slightly erratic one. At the equator the length of day equals the length of night the year through. At the poles there are six months day and six months night, and at intermediate latitudes the time of sunrise and of sunset varies with the season. Not only does the time of sunrise vary, but the time it takes the sun apparently to swing once around the earth also varies. Thus from noon by the sun until noon by the sun again is sometimes more than twenty-four hours and sometimes less than twenty-four hours. The reasons for this variation will be taken up in the chapter on the earth's revolution.

Mean Solar Day. By a mean solar day is meant the average interval of time from sun noon to sun noon. While the apparent solar day varies, the mean solar day is exactly twenty-four hours long. A sundial does not record the same time as a clock, as a usual thing, for the sundial records apparent solar time while the clock records mean solar time.

Relation of Longitude to Time. The sun's apparent daily journey around the earth with the other bodies of the celestial sphere gives rise to day and night* It takes the sun, on the average, twenty-four hours apparently to swing once around the earth. In this daily journey it crosses 360° of longitude, or 15° for each hour. It thus takes four minutes for the sun's rays to sweep over one degree of longitude. Suppose it is noon by the sun at the 90th meridian, in four minutes the sun will be over the 91st meridian, in four more minutes it will be noon by the sun on the 92d meridian, and so on around the globe.

Students are sometimes confused as to the time of day in places east of a given meridian as compared with the time in places west of it. When the sun is rising here, it has already risen for places east of us, hence their time is after sunrise or later than ours. If it is noon by the sun here, at places east of us, having already been noon there, it must be past noon or later in the day. *Places to the east have later time because the sun reaches them first.* To the westward the converse of this is true. If the sun is rising here, it has not yet risen for places west of us and their time is before sunrise or earlier. When it is noon by the sun in Chicago, the shadow north, it is past noon by the sun in Detroit and other places eastward and before noon by the sun in Minneapolis and other places westward.

How Longitude is Determined. A man when in London, longitude 0°, set his watch according to mean solar time there. When he arrived at home he found the mean solar time to be six hours earlier (or slower) than his watch, which he had not changed. Since his watch indicated later time, London must be east of his home, and since the sun appeared six hours earlier at London, his home must be 6 × 15°, or 90°, west of London. While on shipboard at a certain place he noticed that the sun's shadow was due north when his watch indicated 2.30 o'clock, P.M. Assuming that both

*Many thoughtlessly assume that the fact of day and night is a proof of the earth's rotation.

the watch and the sun were "on time" we readily see that since London time was two and one half hours later than the time at that place, he must have been west of London $2\frac{1}{2} \times 15^\circ$, or 37° 30′.

Ship's Chronometer. Every ocean vessel carries a veryaccurate watch called a chronometer. This is regulated to run as perfectly as possible and is set according to the mean solar time of some well known meridian. Vessels from English speaking nations all have their chronometers set with Greenwich time. By observing the time according to the sun at the place whose longitude is sought and comparing that time with the time of the prime meridian as indicated by the chronometer, the longitude is reckoned. For example, suppose the time according to the sun is found by observation to be 9.30 o'clock, a.m., and the chronometer indicates 11.20 o'clock, a.m. The prime meridian, then, must be east as it has later time. Since the difference in time is one hour and fifty minutes and there are 15° difference in longitude for an hour's difference in time, the difference in longitude must be $1\frac{5}{6} \times 15^\circ$, or 27° 30′.

The relation of longitude and time should be thoroughly mastered. From the table at the close of this chapter, giving the longitude, of the principal cities of the world, one can determine the time it is in those places when it is noon at home. Many other problems may also be suggested. It should be borne in mind that it is the *mean solar time* that is thus considered, which in most cities is not the time indicated by the watches and clocks there. People all over Great Britain set their timepieces to agree with Greenwich time, in Ireland with Dublin, in France with Paris, etc. (see "Time used in Various Countries" at the end of this chapter).

Local Time. The mean solar time of any place is often called its local time. This is the average time indicated by the sundial. All places on the same meridian have the same local time. Places on different meridians must of necessity have different local time, the difference in time being four minutes for every degree's difference in longitude.

STANDARD TIME

Origin of Present System. Before the year 1883, the people of different cities in the United States commonly used the local time of the meridian passing through the city. Prior to the advent of the railroad, telegraph, and telephone, little inconvenience was occasioned by the prevalence of so many time systems. But as transportation and communication became rapid and complex it became very difficult to adjust one's time and calculations according to so many standards as came to prevail. Each railroad had its own arbitrary system of time, and where there were several railroads in a city there were usually as many species of "railroad time" besides the local time according to longitude.

"Before the adoption of standard time there were sometimes as many as five different kinds of time in use in a single town. The railroads of the United States followed fifty-three different standards, whereas they now use five. The times were very much out of joint."*

His inability to make some meteorological calculations in 1874 because of the diverse and doubtful character of the time of the available weather reports, led Professor Cleveland Abbe, for so many years connected with the United States Weather service, to suggest that a system of standard time should be adopted. At about the same time several others made similar suggestions and the subject was soon taken

*The Scrap Book, May, 1906.

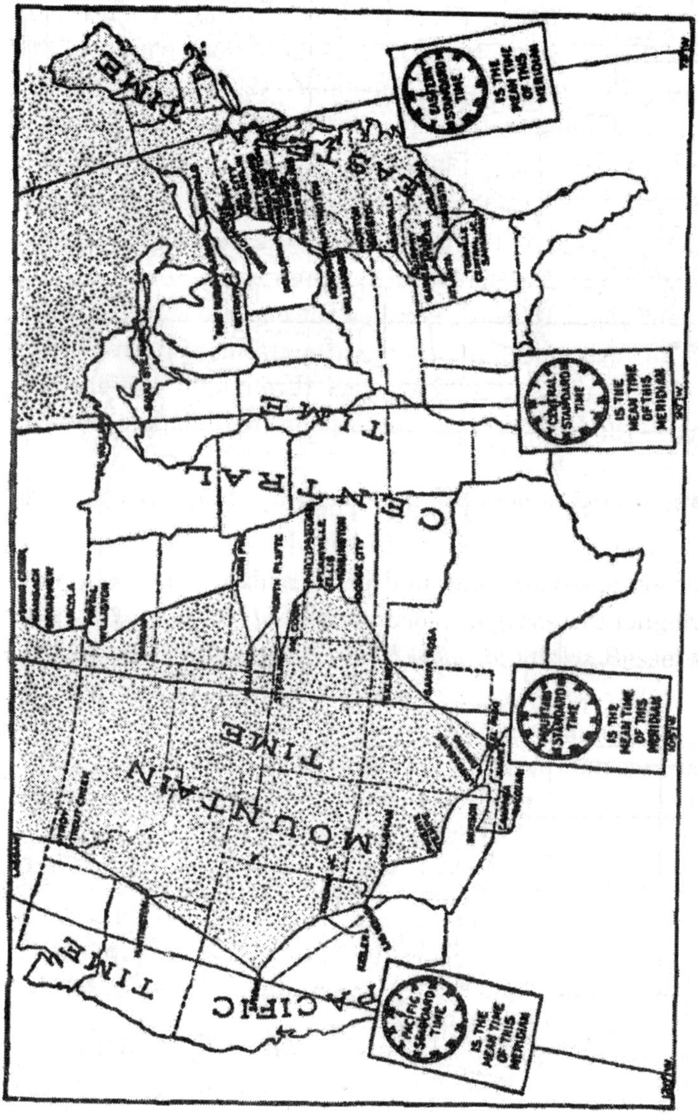

Fig. 27. Standard Time Belts

up in an official way by the railroads of the country under the leadership of William F. Allen, then secretary of the General Time Convention of Railroad Officials. As a result of his untiring efforts the railway associations endorsed his plan and at noon of Sunday, November 18, 1883, the present system was inaugurated.

Eastern Standard Time. According to the system all cities approximately within 7½° of the 75th meridian use the mean solar time of that meridian, the clocks and watches being thus just five hours earlier than those of Greenwich. This belt, about 15° wide, is called the eastern standard time belt. The 75th meridian passes through the eastern portion of Philadelphia, so the time used throughout the eastern portion of the United States corresponds to Philadelphia local mean solar time.

Central Standard Time. The time of the next belt is the mean solar time of the 90th meridian or one hour slower than eastern standard time. This meridian passes through or very near Madison, Wisconsin, St. Louis, and New Orleans, where mean local time is the same as standard time. When it is noon at Washington, D. C, it is 11 o'clock, a.m., at Chicago, because the people of the former city use eastern standard time and those at the latter use central standard time.

Mountain Standard Time. To the west of the central standard time belt lies the mountain region where the time used is the mean solar time of the 105th meridian. This meridian passes through Denver, Colorado, and its clocks as a consequence indicate the same time that the mean sun does there. As the standard time map shows, all the belts are bounded by irregular lines, due to the fact that the people of a city usually use the same time that their principal railroads do, and where trains change their time depends in a large measure upon the convenience to be served. This belt shows the anomaly of being bounded on the east by the central time belt, on the west by the Pacific time belt, and on the *south* by the same belts. The reasons why the mountain standard time belt tapers to a point at the south and the peculiar conditions which consequently result, are discussed under the topic "Four Kinds of Time around El Paso" (p. 74).

Pacific Standard Time. People living in the states bordering or near the Pacific Ocean use the mean solar time of the 120th meridian and thus have three hours earlier time than the people of the Atlantic coast states. This meridian forms a portion of the eastern boundary of California.

In these great time belts* all the clocks and other timepieces differ in time by whole hours. In addition to astronomical observatory clocks, which are regulated according to the mean local time of the meridian passing through the observatory, there are a few cities in Michigan, Georgia, New Mexico, and elsewhere in the United States, where mean local time is still used.

Standard Time in Europe. In many European countries standard time based upon Greenwich time, or whole hour changes from it, is in general use, although there are many more cities which use mean local time than in the United States. *Western European time*, or that of the meridian of Greenwich, is used in Great Britain, Spain, Belgium, and Holland. *Central European time*, one hour later than that of Greenwich, is used in Norway, Sweden, Denmark, Luxemburg, Germany, Switzerland, Austria-Hungary, Servia, and Italy. *Eastern European time*, two hours later than that of Greenwich, is used in Turkey, Bulgaria, and Roumania.

*For a discussion of the time used in other portions of North America and elsewhere in the world see pp.80–87.

TELEGRAPHIC TIME SIGNALS

Getting the Time. An admirable system of sending time signals all over the country and even to Alaska, Cuba, and Panama, is in vogue in the United States, having been established in August, 1865. The Naval Observatories at Washington, D. C., and Mare Island, California, send out the signals during the five minutes preceding noon each day.

It is a common custom for astronomical observatories to correct their own clocks by careful observations of the stars. The Washington Observatory sends telegraphic signals to all the cities east of the Rocky Mountains and the Mare Island Observatory to Pacific cities and Alaska. A few railroads receive their time corrections from other observatories. Goodsell Observatory, Carleton College, Northfield, Minnesota, has for many years furnished time to the Great Northern, the Northern Pacific, the Great Western, and the Sault Ste. Marie railway systems. Allegheny Observatory sends out time to the Pennsylvania system and the Lick Observatory to the Southern Pacific system.

How Time is Determined at the United States Naval Observatory. The general plan of correcting clocks at the United States Naval Observatories by stellar observations is as follows: A telescope called a meridian transit is situated in a true north-south direction mounted on an east-west axis so that it can be rotated in the plane of the meridian but not in the slightest degree to the east or west. Other instruments used are the chronograph and the sidereal clock. The chronograph is an instrument which may be electrically connected with the clock and which automatically makes a mark for each second on a sheet of paper fastened to a cylinder. The sidereal clock is regulated to keep time with the stars—not with the sun, as are other clocks. The reason for this is because the stars make an apparent circuit with each rotation of the earth and this, we have observed, is unerring while the sun's apparent motion is quite irregular.

To correct the clock, an equatorial or high zenith star is selected. A well known one is chosen since the exact time it will cross the meridian of the observer (that is, be at its highest point in its apparent daily rotation) must be calculated. The chronograph is then started, its pen and ink adjusted, and its electrical wires connected with the clock. The observer now sights the telescope to the point where the expected star will cross his meridian and, with his hand on the key, he awaits the appearance of the star. As the star crosses each of the eleven hair lines in the field of the telescope, the observer presses the key which automatically marks upon the chronographic cylinder. Then by examining the sheet he can tell at what time, *by the clock*, the star crossed the center line. He then calculates just what time the clock *should* indicate and the difference is the error of the clock. By this means an error of one tenth of a second can be discovered.

The Sidereal Clock. The following facts concerning the sidereal clock may be of interest. It is marked with twenty-four hour spaces instead of twelve. Only one moment in the year does it indicate the same time as ordinary timepieces, which are adjusted to the average sun. When the error of the clock is discovered the clock is not at once reset because any tampering with the clock would involve a slight error. The correction is simply noted and the rate of the clock's gaining or losing time is calculated so that the true time can be ascertained very precisely at any time by referring to the data showing the clock error when last corrected and the rate at which it varies.

A while before noon each day the exact sidereal time is calculated; this is converted into local mean solar time and this into standard time. The Washington Naval Observatory converts this into the standard time of the 75th meridian or Eastern time and the Mare Island Observatory into that of the 120th meridian or Pacific time.

Sending Time Signals. By the co-operation of telegraph companies, the time signals which are sent out daily from the United States Naval Observatories reach practically every telegraph station in the country. They are sent out at noon, 75th meridian time, from Washington, which is 11 o'clock, a.m., in cities using Central time and 10 o'clock, a.m., where Mountain time prevails; and at noon, 120th meridian time, they are sent to Pacific coast cities from the Mare Island Observatory—three hours after Washington has ashed the signal which makes correct time accessible to sixty millions of our population living east of the Rockies.

Not only are the time signals sent to the telegraph stations and thence to railway offices, clock makers and repairers, schools, court houses, etc., but the same telegraphic signal that marks noon also actually sets many thousands of clocks, their hands whether fast or slow automatically flying to the true mark in response to the electric current. In a number of cities of the United States, nineteen at present, huge balls are placed upon towers or buildings and are automatically dropped by the electric noon signal. The time ball in Washington is conspicuously placed on the top of the State, War, and Navy building and may be seen at considerable distances from many parts of the city.

A few minutes before noon each day, one wire at each telegraphic office is cleared of all business and "thousands of telegraph operators sit in silence, waiting for the click of the key which shall tell them that the 'master clock' in

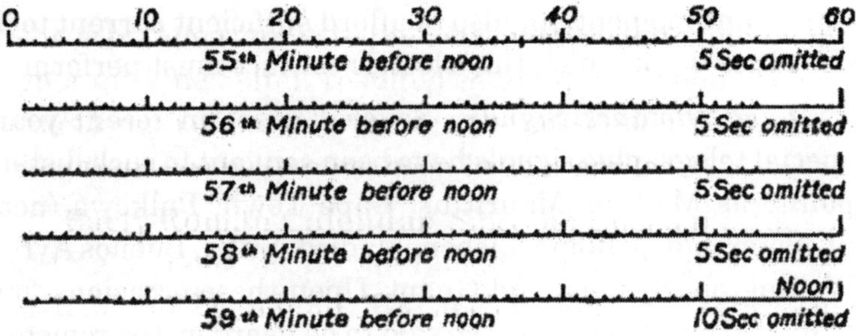

Fig. 28

Washington has begun to speak."* At five minutes before twelve the instrument begins to click off the seconds. Figure 28 (adapted from a cut appearing in Vol. IV, Appendix IV, United States Naval Observatory Publications) graphically shows which second beats are sent along the wires during each of the five minutes before noon by the transmitting clock at the Naval Observatory.

Explanation of the Second Beats. It will be noticed that the twenty-ninth second of each minute is omitted. This is for the purpose of permitting the observer to distinguish, without counting the beats, which is the one denoting the middle of each minute; the five seconds at the end of each of the first four minutes are omitted to mark the beginning of a new minute and the last ten seconds of the fifty-ninth minute are omitted to mark conspicuously the moment of noon. The omission of the last ten seconds also enables the operator to connect his wire with the clock to be automatically set or the time ball to be dropped. The contact marking noon is prolonged a full second, not only to make prominent this

*From "What's the Time," Youth's Companion, May 17 and June 14, 1906.

important moment but also to afford sufficient current to do the other work which this electric contact must perform.

Long Distance Signals. Several times in recent years special telegraphic signals have been sent out to such distant points as Madras, Mauritius, Cape Town, Pulkowa (near St. Petersburg), Rome, Lisbon, Madrid, Sitka, Buenos Ayres, Wellington, Sydney, and Guam. Upon these occasions "our standard clock may fairly be said to be heard in 'the remotest ends of the earth,' thus anticipating the day when wireless telegraphy will perhaps allow of a daily international time signal that will reach every continent and ocean in a small fraction of a second."*

These reports have been received at widely separated stations within a few seconds, being received at the Lick Observatory in 0.05^s, Manila in 0.11^s, Greenwich in 1.33^s, and Sydney, Australia, in 2.25^s.

CONFUSION FROM VARIOUS STANDARDS

Where different time systems are used in the same community, confusion must of necessity result. The following editorial comment in the *Official Railway Guide* for November, 1900, very succinctly sets forth the condition which prevailed in Detroit as regards standard and local time.

"The city of Detroit is now passing through an agitation which is a reminiscence of those which took place throughout the country about seventeen years ago, when standard time was first adopted. For some reason, which it is difficult to explain, the city fathers of Detroit have refused to change from the old local time to the standard, notwithstanding the

*"The Present Status of the Use of Standard Time," by Lieut. Commander E. E. Hayden, U. S. Navy.

fact that all of the neighboring cities—Cleveland, Toledo, Columbus, Cincinnati, etc.,—in practically the same longitude, had made the change years ago and realized the benefits of so doing. The business men of Detroit and visitors to that city have been for a long time laboring under many disadvantages owing to the confusion of standards, and they have at last taken the matter into their own hands and a lively campaign, with the co-operation of the newspapers, has been organized during the past two months. Many of the hotels have adopted standard time, regardless of the city, and the authorities of Wayne County, in which Detroit is situated, have also decided to hold court on Central Standard time, as that is the official standard of the state of Michigan. The authorities of the city have so far not taken action. It is announced in the newspapers that they probably will do so after the election, and by that time, if progress continues to be made, the only clock in town keeping the local time will be on the town hall. All other matters will be regulated by standard time, and the hours of work will have been altered accordingly in factories, stores, and schools. Some opposition has been encountered, but this, as has been the case in every city where the change has been made, comes from people who evidently do not comprehend the effects of the change. One individual, for instance, writes to a newspaper that he will decline to pay pew rent in any church whose clock tower shows standard time; he refuses to have his hours of rest curtailed. How these will be affected by the change he does not explain. Every visitor to Detroit who has encountered the difficulties which the confusion of standards there gives rise to, will rejoice when the complete change is effected."

The longitude of Detroit being 83° W., it is seven degrees east of the 90th meridian, hence the local time used in the city was twenty-eight minutes faster than Central time and thirty-two minutes slower than Eastern time. In Gainesville, Georgia, mean local sun time is used in the city, while the Southern railway passing through the city uses Eastern time and the Georgia railway uses Central time.

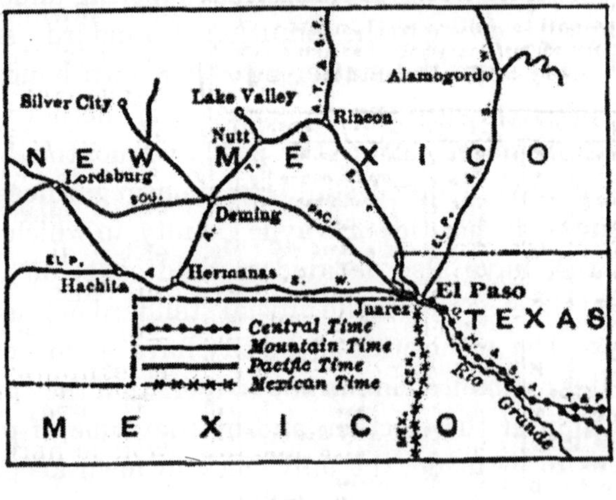

Fig. 29

Four Kinds of Time Around El Paso. Another place of peculiar interest in connection with this subject is El Paso, Texas, from the fact that four different systems are employed. The city, the Atchison, Topeka, and Santa Fe, and the El Paso and Southwestern railways use Mountain time. The Galveston, Harrisburg, and San Antonio, and the Texas and Pacific railways use Central time. The Southern Pacific railway uses Pacific time. The Mexican Central railway uses Mexican standard time. From this it will be seen that when clocks in Strauss, N. M., a few miles from El Paso, are striking twelve, the clocks in El Paso are striking one; in Ysleta, a few miles east, they are striking two; while across the river in Juarez, Mexico, the clocks indicate 12.24.

Time Confusion for Travelers. The confusion which prevails where several different standards of time obtain is well illustrated in the following extract from "The Impressions of a Careless Traveler" by Lyman Abbott, in the Outlook, Feb. 28, 1903.

"The changes in time are almost as interesting and quite as perplexing as the changes in currency. Of course our steamer time changes every day; a sharp blast on the whistle notifies us when it is twelve o'clock, and certain of the passengers set their watches accordingly every day. I have too much respect for my faithful friend to meddle with him to this extent, and I keep my watch unchanged and make my calculations by a mental comparison of my watch with the ship's time. But when we are in port we generally have three times—ship's time, local time, and railroad time, to which I must in my own case add my own time, which is quite frequently neither. In fact, I kept New York time till we reached Genoa; since then I have kept central Europe railroad time. Without changing my watch, I am getting back to that standard again, and expect to find myself quite accurate when we land in Naples."

LEGAL ASPECT OF STANDARD TIME

The legal aspect of standard time presents many interesting features. Laws have been enacted in many different countries and several of the states of this country legalizing some standard of time. Thus in Michigan, Minnesota, and other central states the legal time is the mean solar time of longitude 90° west of Greenwich. When no other standard is explicitly referred to, the time of the central belt is the legal time in force. Similarly, legal time in Germany was declared by an imperial decree dated March 12, 1903, as follows:*

*Several of the following quotations are taken from the "Present Status of the Use of Standard Time," by E. E. Hayden.

"We, Wilhelm, by the grace of God German Emperor, King of Prussia, decree in the name of the Empire, the Bundesrath and Reichstag concurring, as follows:

"The legal time in Germany is the mean solar time of longitude 15° east from Greenwich."

Greenwich time for Great Britain, and Dublin time for Ireland, were legalized by an act of Parliament as follows:

A Bill to remove doubts as to the meaning of expressions relative to time occurring in acts of Parliament, deeds, and other legal instruments.

Whereas it is expedient to remove certain doubts as to whether expressions of time occurring in acts of Parliament, deeds, and other legal instruments relate in England and Scotland to Greenwich time, and in Ireland to Dublin time, or to the mean astronomical time in each locality:

Be it therefore enacted by the Queen's most Excellent Majesty, by and with the advice and consent of the Lords, spiritual and temporal, and Commons in the present Parliament assembled, and by the authority of the same, as follows (that is to say):

1. That whenever any expression of time occurs in any act of Parliament, deed, or other legal instrument, the time referred to shall, unless it is otherwise specifically stated, be held in the case of Great Britain to be Greenwich mean time and in the case of Ireland, Dublin mean time.

2. This act may be cited as the statutes (definition of time) act, 1880.

Seventy-fifth meridian time was legalized in the District of Columbia by the following act of Congress:

An Act to establish a standard of time in the District of Columbia. Be it enacted by the Senate and House of Representatives of the United States of America in Congress assembled, That the legal standard of time in the District of Columbia shall hereafter be the mean time of the seventy-fifth meridian of longitude west from Greenwich.

Section 2. That this act shall not be so construed as to affect existing contracts.

Approved, March 13, 1884.

In New York eastern standard time is legalized in section 28 of the Statutory Construction Law as follows:

The standard time throughout this State is that of the 75th meridian of longitude west from Greenwich, and all courts and public offices, and legal and official proceedings, shall be regulated thereby. Any act required by or in pursuance of law to be performed at or within a prescribed time, shall be performed according to such standard time.

A New Jersey statute provides that the time of the same meridian shall be that recognized in all the courts and public offices of the State, and also that "the time named in any notice, advertisement, or contract shall be deemed and taken to be the said standard time, unless it be otherwise expressed." In Pennsylvania also it is provided that "on and after July 1, 1887, the mean solar time of the seventy-fifth meridian of longitude west of Greenwich, commonly called eastern standard time," shall be the standard in all public matters; it is further provided that the time "in any and all contracts, deeds, wills, and notices, and in the transaction of all matters of business, public, legal, commercial, or otherwise, shall be construed with reference to and in accordance with the said standard hereby adopted, unless a different standard is therein expressly provided for."

Where there is no standard adopted by legal authority, difficulties may arise, as the following clipping from the *New York Sun*, November 25, 1902, illustrates:

WHAT'S NOON IN A FIRE POLICY?

Solar Noon or Standard Time Noon—Courts Asked to Say.

Fire in Louisville at 11:45 a.m., Standard Time, Which Was 12:02 1-2 p.m. Solar Time—Policies Expired at Noon and 13 Insurance Companies Wont Pay.

Whether the word "noon," which marks the beginning and expiration of all fire insurance policies, means noon by standard time, or noon by solar time, is a question which is soon to be fought out in the courts of Kentucky, in thirteen suits which have attracted the attention of fire insurance people all over the world. The suits are being brought by the Peaslee-Gaulbert Company and the Louisville Lead and Color Company of Louisville, and $19,940.70 of insurance money depends on the result.

Now, although the policies in these companies all state that they were in force from noon of April 1, 1901, to noon of April 1, 1902, not one of them says what kind of time that period of the day is to be reckoned in. In Louisville the solar noon is 17½ minutes earlier than the standard noon, so that a fire occurring in the neighborhood of noon on the day of a policy's expiration, may easily be open to attack.

The records of the Louisville fire department show that the fire that destroyed the buildings of the two companies was discovered at 11.45 o'clock Louisville standard time in the forenoon of April 1, last. The fire began in the engine room of the main factory and spread to the two other buildings which were used mainly as warehouses. When the fire department recorded the time of the fire's discovery it figured, of course, by standard time. Solar time would make it just two and a half minutes after noon. If noon in the policies means noon by solar time, of course the thirteen companies are absolved from any responsibility for the loss. If it means noon by standard time, of course they must pay.

When the insurance people got the claims of the companies they declined to pay, and when asked for an explanation declared that noon in the policies meant noon by solar time. The burned-out companies immediately began suit, and in their affidavits they say that not only is standard time the official time of the State of Kentucky and the city of Louisville, but it is also the time upon which all business engagements and all domestic and social engagements are reckoned. They state further that they are prepared to show that in 1890 the city of Louisville passed an ordinance making standard time the official time of the city, that all legislation is dated according to standard time, and that the governor of the state is inaugurated at noon according to the same measurement of time.

Solar time, state the companies, can be found in use in Louisville by only a few banking institutions which got charters many years ago that compel them to use solar time to this day. Most banks, they say, operate on standard time, although they keep clocks going at solar time so as to do business on that basis if requested. Judging by standard time the plainti"s allege their fire took place fifteen minutes before their policies expired.

The suits will soon come to trial, and, of course, will be watched with great interest by insurance people.

Iowa Case. An almost precisely similar case occurred at Creston, Iowa, September 19, 1897. In this instance the insurance policies expired "at 12 o'clock at noon," and the fire broke out at two and a half minutes past noon according to standard time, but at fifteen and one-half minutes before local mean solar noon. In each of these cases the question of whether standard time or local mean solar time was the accepted meaning of the term was submitted to a jury, and in the first instance the verdict was in favor of standard time, in the Iowa case the verdict was in favor of local time.

Early Decision in England. In 1858 and thus prior to the formal adoption of standard time in Great Britain, it was held that the time appointed for the sitting of a court must be understood as the mean solar time of the place where the court is held and not Greenwich time, unless it be so expressed, and a new trial was granted to a defendant who had arrived at the local time appointed by the court but found the court had met by Greenwich time and the case had been decided against him.

Court Decision in Georgia. In a similar manner a court in the state of Georgia rendered the following opinion:

> "The only standard of time in computation of a day, or hours of a day, recognized by the laws of Georgia is the meridian of the sun; and a legal day begins and ends at midnight, the mean time between meridian and meridian, or 12 o'clock *post meridiem*. An arbitrary and artificial standard of time, fixed by persons in a certain line of business, cannot be substituted at will in a certain locality for the standard recognized by the law."

Need for Legal Time Adoption on a Scientific Basis. There is nothing in the foregoing decisions to determine whether mean local time, or the time as actually indicated by the sun at a particular day, is meant. Since the latter sometimes varies as much as fifteen minutes faster or slower than the average, opportunities for controversies are multiplied when no scientifically accurate standard time is adopted by law.

Even though statutes are explicit in the definition of time, they are still subject to the official interpretation of the courts, as the following extracts show:

Thomas Mier took out a fire insurance policy on his saloon at 11.30 standard time, the policy being dated noon of that day. At the very minute that he was getting the policy the saloon caught fire and was burned. Ohio law makes standard time legal time, and the company refused to pay the $2,000 insurance on Mier's saloon. The case was fought through to the Supreme Court, which decided that "noon" meant the time the sun passed the meridian at Akron, which is at 11.27 standard time. The court ordered the insurance company to pay.—*Law Notes*, June, 1902.

In the 28th Nebraska Reports a case is cited in which judgment by default was entered against a defendant in a magistrate's court who failed to make an appearance at the stipulated hour by standard time, but arrived within the limit by solar time. He contested the ruling of the court, and the supreme judiciary of the state upheld him in the contest, although there was a Nebraska statute making standard time the legal time. The court held that "at noon" must necessarily mean when the sun is over the meridian, and that no construction could reasonably interpret it as indicating 12 o'clock standard time.

TIME USED IN VARIOUS COUNTRIES

The following table is taken, by permission, largely from the abstracts of official reports given in Vol. IV, Appendix IV of the Publications of the United States Naval Observatory, 1905. The time given is fast or slow as compared with Greenwich mean solar time.

Argentina, 4 h. 16 m. 48.2 s. slow. Official time is referred to the meridian of Cordoba. At 11 o'clock, a.m., a daily signal is telegraphed from the Cordoba Observatory.

Austria-Hungary, 1 h. fast. Standard time does not exist except for the service of railroads where it is in force, not by law, but by order of the proper authorities.

Belgium. Official time is calculated from 0 to 24 hours, zero corresponding to midnight at Greenwich. The Royal Observatory at Brussels communicates daily the precise hour by telegraph.

British Empire.

Great Britain. The meridian of Greenwich is the standard time meridian for England, Isle of Man, Orkneys, Shetland Islands, and Scotland.

Ireland, 0 h. 25 m. 21.1 s. slow. The meridian of Dublin is the standard time meridian.

Africa (English Colonies), 2 h. fast. Standard time for Cape Colony, Natal, Orange River Colony, Rhodesia and Transvaal.

Australia.

New South Wales, Queensland, Tasmania and Victoria, 10 h. fast.

South Australia and Northern Territory, 9 h. 30 m. fast.

Canada.

Alberta and Saskatchewan, 7 h. slow.

British Columbia, 8 h. slow.

Keewatin and Manitoba, 6 h. slow.

Ontario and Quebec, 5 h. slow.

New Brunswick, Nova Scotia, and Prince Edward Island, 4 h. slow.

Chatham Island, 11 h. 30 m. fast.

Gibraltar, Greenwich time.

Hongkong, 8 h. fast.

Malta, 1 h. fast.

New Zealand, 11 h. 30 m. fast.

India. Local mean time of the Madras Observatory, 5 h. 20 m. 59.1 s., is practically used as standard time for India and Ceylon, being telegraphed daily all over the country; but for strictly local use it is generally converted into local mean time. It is proposed soon to adopt the standard time of 5 h. 30 m. fast of Greenwich for India and Ceylon, and 6 h. 30 m. fast of Greenwich for Burmah.

Newfoundland, 3 h. 30 m. 43.6 s. slow. (Local mean time of St. John's.)

Chile, 4 h. 42 m. 46.1 s. slow. The official railroad time is furnished by the Santiago Observatory. It is telegraphed over the country daily at 7 o'clock, a.m. The city of Valparaiso uses the local time, 4 h. 46 m. 34.1 s. slow, of the observatory at the Naval School located there.

China. An observatory is maintained by the Jesuit mission at Zikawei near Shanghai, and a time ball suspended from a mast on the French Bund in Shanghai is dropped electrically precisely at noon each day. This furnishes the local time at the port of Shanghai 8 h. 5 m. 43.3 s. fast, which is adopted by the railway and telegraph companies represented there, as well as by the coastwise shipping. From Shanghai the time is telegraphed to other ports. The Imperial Railways of North China use the same time, taking it from the British gun at Tientsin and passing it on to the stations of the railway twice each day, at 8 o'clock a.m. and at 8 o'clock p.m. Standard time, 7 h. and 8 h. fast, is coming into use all along the east coast of China from Newchwang to Hongkong.

Colombia. Local mean time is used at Bogota, 4 h. 56 m. 54.2 s. slow, taken every day at noon in the observatory. The lack of effective telegraphic service makes it impossible to communicate the time as corrected at Bogota to other parts of the country, it frequently taking four and five days to send messages a distance of from 50 to 100 miles.

Costa Rica, 5 h. 36 m. 16.9 s. slow. This is the local mean time of the Government Observatory at San Jos□e.

Cuba, 5 h. 29 m. 26 s. slow. The official time of the Republic is the civil mean time of the meridian of Havana and is used by the railroads and telegraph lines of the government. The Central Meteorological Station gives the time daily to the port and city of Havana as well as to all the telegraph offices of the Republic.

Denmark, 1 h. fast. In Iceland, the Faroe Islands and the Danish West Indies, local mean time is used.

Egypt, 2 h. fast. Standard time is sent out electrically by the standard clock of the observatory to the citadel at Cairo, to Alexandria, Port Said and Wady-Halfa.

Equador, 5 h. 14 m. 6.7 s. slow. The official time is that of the meridian of Quito, corrected daily from the National Observatory.

France, 0 h. 9 m. 20.9 s. fast. Legal time in France, Algeria and Tunis is local mean time of the Paris Observatory. Local mean time is considered legal in other French colonies.

German Empire.

Germany, 1 h. fast.

Kiaochau, 8 h. fast.

Southwest Africa, 1 h. fast.

It is proposed to adopt standard time for the following:

Bismarck Archipelago, Carolines, Mariane Islands and New Guinea, 10 h. fast.

German East Africa, 2 h. fast or 2 h. 30 m. fast.

Kamerun, 1 h. fast.

Samoa (after an understanding with the U. S.), 12 h. fast.

Toga, Greenwich time.

Greece, 1 h. 34 m. 52.9 s. fast. By royal decree of September 14, 1895, the time in common use is that of the mean time of Athens, which is transmitted from the observatory by telegraph to the towns of the kingdom.

Holland. The local time of Amsterdam, 0 h. 19 m. 32.3 s. fast, is generally used, but Greenwich time is used by the post and telegraph administration and the railways and other transportation companies. The observatory at Leyden communicates the time twice a week to Amsterdam, The Hague, Rotterdam and other cities, and the telegraph bureau at Amsterdam signals the time to all the other telegraph bureaus every morning.

Honduras. In Honduras the half hour nearest to the meridian of Tegucigalpa, longitude 87°12' west from Greenwich, is generally used. Said hour, 6 h. slow, is frequently determined at the National Institute by means of a solar chronometer and communicated by telephone to the Industrial School, where in turn it is indicated to the public by a steam whistle. The central telegraph office communicates it to the various sub-offices of the Republic, whose clocks generally serve as a basis for the time of the villages, and in this manner an approximately uniform time is established throughout the Republic.

Italy, 1 h. fast. Adopted by royal decree of August 10, 1893. This time is kept in all government establishments, ships of the Italian Navy in the ports of Italy, railroads, telegraph offi ces, and Italian coasting steamers. The hours are numbered from 0 to 24, beginning with midnight.

Japan. Imperial ordinance No. 51, of 1886: "The meridian that passes through the observatory at Greenwich, England, shall be the zero (0) meridian. Longitude shall be counted from the above meridian east and west up to 180 degrees, the east being positive and the west negative. From January 1, 1888, the time of the 135th degree east longitude shall be the standard time of Japan." This is 9 h. fast. Imperial ordinance No. 167, of 1895: "The standard time hitherto used in Japan shall henceforth be called central standard time. The time of the 120th degree east longitude shall be the standard time of Formosa, the Pescadores, the Yaeyama, and the

Miyako groups, and shall be called western standard time. This ordinance shall take effect from the first of January, 1896." This is 8 h. fast.

Korea, 8 h. 30 m. fast. Central standard time of Japan is telegraphed daily to the Imperial Japanese Post and Telegraph Office at Seoul. Before December, 1904, this was corrected by subtracting 30 m., which nearly represents the difference in longitude, and was then used by the railroads, street railways, and post and telegraph offices, and most of the better classes. Since December 1, 1904, the Japanese post-offices and railways in Korea have begun to use central standard time of Japan. In the country districts the people use sundials to some extent.

Luxemburg, 1 h. fast, the legal and uniform time.

Mexico, 6 h. 36 m. 26.7 s. slow. The National Astronomical Observatory of Tacubaya regulates a clock twice a day which marks the local mean time of the City of Mexico, and a signal is raised twice a week at noon upon the roof of the national palace, such signal being used to regulate the city's public clocks. This signal, the clock at the central telegraph office, and the public clock on the cathedral, serve as a basis for the time used commonly by the people. The general telegraph office transmits this time daily to all of its branch offices. Not every city in the country uses this time, however, since a local time, very imperfectly determined, is more commonly observed. The following railroad companies use standard City of Mexico time corrected daily by telegraph: Central, Hidalgo, Xico and San Rafael, National and Mexican. The Central and National railroads correct their clocks to City of Mexico time daily by means of the noon signal sent out from the Naval Observatory at Washington (see page 70) and by a similar signal from the observatory at St. Louis, Missouri. The Nacozari, and the Cananea, Yaqui River and Pacific railroads use Mountain time, 7 h. slow, and the Sonora railroad uses the local time of Guaymas, 7 h. 24 m. slow.

Nicaragua, 5 h. 45 m. 10 s. slow. Managua time is issued to all public offices, railways, telegraph offices and churches in a zone that extends from San Juan del Sur, latitude 11° 15' 44'' N., to El Ocotal, latitude 12° 46' N., and from El Castillo, longitude 84° 22' 37'' W., to Corinto, longitude 87° 12' 31'' W. The time of the Atlantic ports is usually obtained from the captains of ships.

Norway, 1 h. fast. Central European time is used everywhere throughout the country. Telegraphic time signals are sent out once a week to the telegraph stations throughout the country from the observatory of the Christiania University.

Panama. Both the local mean time of Colon, 5 h. 19 m. 39 s. slow, and eastern standard time of the United States, 5 h. slow, are used. The latter time is cabled daily by the Central and South American Cable Company from the Naval Observatory at Washington, and will probably soon be adopted as standard.

Peru, 5 h. 9 m. 3 s. slow. There is no official time. The railroad from Callao to Oroya takes its time by telegraph from the noon signal at the naval school at Callao, which may be said to be the standard time for Callao, Lima, and the whole of central Peru. The railroad from Mollendo to Lake Titicaca, in southern Peru, takes its time from ships in the Bay of Mollendo.

Portugal, 0 h. 36 m. 44.7 s. slow. Standard time is in use throughout Portugal and is based upon the meridian of Lisbon. It is established by the Royal Observatory in the Royal Park at Lisbon, and from there sent by telegraph to every railway station throughout Portugal having telegraphic communication. Clocks on railway station platforms are five minutes behind and clocks outside of stations are true.

Russia, 2 h. 1 m. 18.6 s. fast. All telegraph stations use the time of the Royal Observatory at Pulkowa, near St. Petersburg. At railroad stations both local and Pulkowa time are given, from which it is to be inferred that for all local purposes local time is used.

Salvador, 5 h. 56 m. 32 s. slow. The government has established a national observatory at San Salvador which issues time on Wednesdays and Saturdays, at noon, to all public offices, telegraph offices, railways, etc., throughout the Republic.

Santo Domingo, 4 h. 39 m. 32 s. slow. Local mean time is used, but there is no central observatory and no means of correcting the time. The time differs from that of the naval vessels in these waters by about 30 minutes.

Servia, 1 h. fast. Central European time is used by the railroad, telegraph companies, and people generally. Clocks are regulated by telegraph from Budapest every day at noon.

Spain, Greenwich time. This is the official time for use in governmental offices in Spain and the Balearic Islands, railroad and telegraph offices. The hours are numbered from 0 to 24, beginning with midnight. In some portions local time is still used for private matters.

Sweden, 1 h. fast. Central European time is the standard in general use. It is sent out every week by telegraph from the Stockholm Observatory.

Switzerland, 1 h. fast. Central European time is the only legal time. It is sent out daily by telegraph from the Cantonal Observatory at Neuchatel.

Turkey. Two kinds of time are used, Turkish and Eastern European time, the former for the natives and the latter for Europeans. The railroads generally use both, the latter for the actual running of trains and Turkish time-tables for the benefit of the natives. Standard Turkish time is used generally by the people, sunset being the base, and twelve hours being added for a theoretical sunrise. The official clocks are set daily so as to read 12 o'clock at the theoretical sunrise, from tables showing the times of sunset, but the tower clocks are set only two or three times a week. The government telegraph lines use Turkish time throughout the empire, and St. Sophia time, 1 h. 56 m. 53 s. fast, for telegrams sent out of the country.

United States. Standard time based upon the meridian of Greenwich, varying by whole hours from Greenwich time, is almost universally used, and is sent out daily by telegraph to most of the country, and to Havana and Panama from the Naval Observatory at Washington, and to the Pacific coast from the observatory at Mare Island Navy Yard, California. For further discussions of standard time belts in the United States, see pp. 64–67 and the U. S. standard time belt map. Insular possessions have time as follows:

Porto Rico, 4 h. slow, Atlantic standard time.
Alaska, 9 h. slow, Alaska standard time.
Hawaiian Islands, 10 h. 30 m. slow, Hawaiian standard time.
Guam, 9 h. 30 m. fast, Guam standard time.
Philippine Islands, 8 h. fast, Philippine standard time.
Tutuila, Samoa, 11 h. 30 m. slow, Samoan standard time.

Uruguay, 3 h. 44 m. 48.9 s. slow. The time in common use is the mean time of the meridian of the dome of the Metropolitan Church of Montevideo. The correct time is indicated by a striking clock in the tower of that church. An astronomical geodetic observatory, with meridian telescope and chronometers, has now been established and will in the future furnish the time. It is proposed to install a time ball for the benefit of navigators at the port of Montevideo. An electric time service will be extended throughout the country, using at first the meridian of the church and afterwards that of the national observatory, when constructed.

Venezuela, 4 h. 27 m. 43.6 s. The time is computed daily at the Caracas Observatory from observations of the sun and is occasionally telegraphed to other parts of Venezuela. The cathedral clock at Caracas is corrected by means of these observations. Railway time is at least five minutes later than that indicated by the cathedral clock, which is accepted as standard by the people. Some people take time from the observatory ag, which always falls at noon.

LATITUDE AND LONGITUDE OF CITIES

The latitude and longitude of cities in the following table was compiled from various sources. Where possible, the exact place is given, the abbreviation "O" standing for observatory, "C" for cathedral, etc.

	Latitude	Longitude from Greenwich
Adelaide, S. Australia, Snapper Point	34° 46' 50" S	138° 30' 39" E
Aden, Arabia, Tel. Station......	12° 46' 40" N	44° 58' 58" E
Alexandria, Egypt, Eunos Pt.	31° 11' 43" N	29° 51' 40" E
Amsterdam, Holland, Ch	52° 22' 30" N	4° 53' 04" E
Antwerp, Belgium, O	51° 12' 28" N	4° 24' 44" E
Apia, Samoa, Ruge's Wharf .	13° 48' 56" S	171° 44' 56" W
Athens, Greece, O	37° 58' 21" N	23° 43' 55" E
Bangkok, Siam, Old Br. Fact.	13° 44' 20" N	100° 28' 42" E
Barcelona, Spain, Old Mole Light	41° 22' 10" N	2° 10' 55" E
Batavia, Java, O	6° 07' 40" N	106° 48' 25" E
Bergen, Norway, C	60° 23' 37" N	5° 20' 15" E
Berlin, Germany, O	52° 30' 17" N	13° 23' 44" E
Bombay, India, O	18° 53' 45" N	72° 48' 58" E
Bordeaux, France, O	44° 50' 07" N	00° 31' 23" W
Brussels, Belgium, O	50° 51' 11" N	4° 22' 18" E
Buenos Aires, Custom House	34° 36' 30" S	58° 22' 14" W
Cadiz, Spain, O	36° 27' 40" N	6° 12' 20" W
Cairo, Egypt, O	30° 04' 38" N	31° 17' 14" E

	Latitude	Longitude from Greenwich
Calcutta, Ft. Wm. Semaphore	22° 33' 25" N	88° 20' 11" E
Canton, China, Dutch Light .	23° 06' 35" N	113° 16' 34" E
Cape Horn, South Summit . .	55° 58' 41" S	67° 16' 15" W
Cape Town, S. Africa, O	33° 56' 03" S	18° 28' 40" E
Cayenne, Fr. Guiana, Landing	4° 56' 20" N	52° 20' 25" W
Christiania, Norway, O	59° 54' 44" N	10° 43' 35" E
Constantinople, Turkey, C. . .	41° 00' 16" N	28° 58' 59" E
Copenhagen, Denmark, New O.	55° 41' 14" N	12° 34' 47" E
Dublin, Ireland, O	53° 23' 13" N	6° 20' 30" W
Edinburgh, Scotland, O	55° 57' 23" N	3° 10' 54" W
Florence, Italy, O	43° 46' 04" N	11° 15' 22" E
Gibraltar, Spain, Dock Flag .	36° 07' 10" N	5° 21' 17" W
Glasgow, Scotland, O	55° 52' 43" N	4° 17' 39" W
Hague, The, Holland, Ch	52° 04' 40" N	4° 18' 30" E
Hamburg, Germany, O	53° 33' 07" N	9° 58' 25" E
Havana, Cuba, Morro Lt. H. .	23° 09' 21" N	82° 21' 30" W
Hongkong, China, C	21° 16' 52" N	114° 09' 31" E
Jerusalem, Palestine, Ch	31° 46' 45" N	35° 13' 25" E
Leipzig, Germany, O	51° 20' 06" N	12° 23' 30" E
Lisbon, Portugal, O. (Royal)	38° 42' 31" N	9° 11' 10" W
Liverpool, England, O	53° 24' 04" N	3° 04' 16" W
Madras, India, O	13° 04' 06" N	80° 14' 51" E
Marseilles, France, New O. . .	43° 18' 22" N	5° 23' 43" E
Melbourne, Victoria, O	37° 49' 53" S	144° 58' 32" E
Mexico, Mexico, O	19° 26' 01" N	99° 06' 39" W
Montevideo, Uruguay, C	34° 54' 33" S	56° 12' 15" W
Moscow, Russia, O	55° 45' 20" N	37° 32' 36" E
Munich, Germany, O.	48° 08' 45" N	11° 36' 32" E
Naples, Italy, O	40° 51' 46" N	14° 14' 44" E
Panama, Cent. Am., C	8° 57' 06" N	79° 32' 12" W
Para, Brazil, Custom H	1° 26' 59" S	48° 30' 01" W
Paris, France, O	2° 20' 14" E	48° 50' 11" N
Peking, China	39° 56' 00" N	116° 28' 54" E

	Latitude	Longitude from Greenwich
Pulkowa, Russia, O	59° 46' 19" N	30° 19' 40" E
Rio de Janeiro, Brazil, O.	22° 54' 24" S	43° 10' 21" W
Rome, Italy, O	41° 53' 54" N	12° 28' 40" E
Rotterdam, Holl., Time Ball .	51° 54' 30" N	4° 28' 50" E
St. Petersburg, Russia, *see* Pulkowa		
Stockholm, Sweden, O	59° 20' 35" N	18° 03' 30" E
Sydney, N. S. Wales, O	33° 51' 41" S	151° 12' 23" E
Tokyo, Japan, O	35° 39' 17" N	139° 44' 30" E
Valparaiso, Chile, Light House	33° 01' 30" S	71° 39' 22" W

UNITED STATES

Aberdeen, S. D., N. N. & I. S.	45° 27' 50" N	98° 28' 45" W
Albany, N. Y., New O	42° 39' 13" N	73° 46' 42" W
Ann Arbor, Mich., O.	42° 16' 48" N	83° 43' 48" W
Annapolis, Md., O.	38° 58' 53" N	76° 29' 08" W
Atlanta, Ga., Capitol	33° 45' 19" N	84° 23' 29" W
Attu Island, Alaska, Chichagoff Harbor	52° 56' 01" N	173° 12' 24" E
Augusta, Me., Baptist Ch......	44° 18' 52" N	69° 46' 37" W
Austin, Tex......	32° 00' 40" N	100° 27' 35" W
Baltimore, Md., Wash. Mt......	39° 17' 48" N	76° 36' 59" W
Bangor, Me., Thomas Hill	44° 48' 23" N	68° 46' 59" W
Beloit, Wis., College...........	42° 30' 13" N	89° 01' 46" W
Berkeley, Cal., O.......................	37° 52' 24" N	122° 15' 41" W
Bismarck, N. D.......................	46° 49' 12" N	100° 45' 08" W
Boise, Idaho, Ast. Pier	43° 35' 58" N	116° 13' 04" W
Boston, Mass., State House . .	42° 21' 28" N	71° 03' 50" W
Bu"alo, N. Y............................	42° 53' 03" N	78° 52' 42" W
Charleston, S. C., Lt. House	32° 41' 44" N	79° 52' 58" W
Cheyenne, Wyo., Ast. Sta......	41° 07' 47" N	104° 48' 52" W
Chicago, Ill., O.......................	41° 50' 01" N	87° 36' 36" W
Cincinnati, Ohio	39° 08' 19" N	84° 26' 00" W
Cleveland, Ohio, Lt. H...........	41° 30' 02" N	81° 42' 10" W

	Latitude	Longitude from Greenwich
Columbia, S. C	33° 59' 12" N	81° 00' 12" W
Columbus, Ohio	39° 57' 40" N	82° 59' 40" W
Concord, N. H	43° 11' 48" N	71° 32' 30" W
Deadwood, S. D., P. O	44° 22' 34" N	103° 43' 19" W
Denver, Col., O	39° 40' 36" N	104° 59' 23" W
Des Moines, Iowa	41° 35' 08" N	93° 37' 30" W
Detroit, Mich	42° 20' 00" N	83° 02' 54" W
Duluth, Minn	46° 48' 00" N	92° 06' 10" W
Erie, Pa., Waterworks	42° 07' 53" N	80° 05' 51" W
Fargo, N. D., Agri. College...	46° 52' 04" N	96° 47' 11" W
Galveston, Tex., C	29° 18' 17" N	94° 47' 26" W
Guthrie, Okla	35° 51' 48" N	100° 26' 24" W
Hartford, Conn	41° 45' 59" N	72° 40' 45" W
Helena, Mont	46° 35' 36" N	111° 52' 45" W
Honolulu, Sandwich Islands . .	21° 18' 12" N	157° 51' 34" W
Indianapolis, Ind.......	39° 47' 00" N	86° 05' 00" W
Jackson, Miss	31° 16' 00" N	91° 36' 18" W
Jacksonville, Fla., M. E. Ch. .	30° 19' 43" N	81° 39' 14" W
Kansas City, Mo	39° 06' 08" N	94° 35' 19" W
Key West, Fla., Light House .	24° 32' 58" N	81° 48' 04" W
Lansing, Mich., Capitol	42° 43' 56" N	84° 33' 23" W
Lexington, Ky., Univ.	38° 02' 25" N	84° 30' 21" W
Lincoln, Neb	40° 55' 00" N	96° 52' 00" W
Little Rock, Ark	34° 40' 00" N	92° 12' 00" W
Los Angeles, Cal., Ct. House	34° 03' 05" N	118° 14' 32" W
Louisville, Ky.	38° 15' 08" N	85° 45' 29" W
Lowell, Mass	42° 22' 00" N	71° 04' 00" W
Madison, Wis., O	43° 04' 37" N	89° 24' 27" W
Manila, Luzon, C	14° 35' 31" N	120° 58' 03" E
Memphis, Tenn	35° 08' 38" N	90° 03' 00" W
Milwaukee, Wis., Ct. House .	43° 02' 32" N	87° 54' 18" W
Minneapolis, Minn., O	44° 58' 38" N	93° 14' 02" W
Mitchell, S. D	43° 49' 00" N	98° 00' 14" W
Mobile, Ala., Epis. Church . .	30° 41' 26" N	88° 02' 28" W

	Latitude	Longitude from Greenwich
Montgomery, Ala	32° 22' 46" N	86° 17' 57" W
Nashville, Tenn., O	36° 08' 54" N	86° 48' 00" W
Newark, N. J., M. E. Ch	40° 44' 06" N	74° 10' 12" W
New Haven, Conn., Yale	41° 18' 28" N	72° 55' 45" W
New Orleans, La., Mint	29° 57' 46" N	90° 03' 28" W
New York, N. Y., City Hall . .	40° 42' 44" N	74° 00' 24" W
Northfield, Minn., O	44° 27' 42" N	93° 08' 57" W
Ogden, Utah, O	41° 13' 08" N	111° 59' 45" W
Olympia, Wash	47° 03' 00" N	122° 57' 00" W
Omaha, Neb	41° 16' 50" N	95° 57' 33" W
Pago Pago, Samoa	14° 18' 06" S	170° 42' 31" W
Philadelphia, Pa., State House	39° 56' 53" N	75° 09' 03" W
Pierre, S. D., Capitol	44° 22' 50" N	100° 20' 26" W
Pittsburg, Pa	40° 26' 34" N	80° 02' 38" W
Point Barrow (highest latitude in the United States)	71° 27' 00" N	156° 15' 00" W
Portland, Ore	45° 30' 00" N	122° 40' 30" W
Princeton, N. J., O	40° 20' 58" N	74° 39' 24" W
Providence, R. I., Unit. Ch. .	41° 49' 28" N	71° 24' 20" W
Raleigh, N. C	35° 47' 00" N	78° 40' 00" W
Richmond, Va., Capitol	37° 32' 19" N	77° 27' 02" W
Rochester, N. Y., O	43° 09' 17" N	77° 35' 27" W
Sacramento, Cal	38° 33' 38" N	121° 26' 00" W
St. Louis, Mo	38° 38' 04" N	90° 12' 16" W
St. Paul, Minn	44° 52' 56" N	93° 05' 00" W
San Francisco, Cal., C. S. Sta	37° 47' 55" N	122° 24' 32" W
San Juan, Porto Rico, Morro Light House	18° 28' 56" N	66° 07' 28" W
Santa Fe, N. M	35° 41' 19" N	105° 56' 45" W
Savannah, Ga., Exchange	32° 04' 52" N	81° 05' 26" W
Seattle, Wash., C. S. Ast. Sta	47° 35' 54" N	122° 19' 59" W
Sitka, Alaska, Parade Ground	57° 02' 52" N	135° 19' 31" W
Tallahassee, Fla	30° 25' 00" N	84° 18' 00" W
Trenton, N. J. Capitol	40° 13' 14" N	74° 46' 13" W

	Latitude	Longitude from Greenwich
Virginia City, Nev	39° 17' 36" N	119° 39' 06" W
Washington, D. C., O	38° 53' 39" N	77° 03' 06" W
Wheeling, W. Va	40° 05' 16" N	80° 44' 30" W
Wilmington, Del., Town Hall	39° 44' 27" N	75° 33' 03" W
Winona, Minn	44° 04' 00" N	91° 30' 00" W

CHAPTER-V

CIRCUMNAVIGATION AND TIME

Magellan's Fleet. When the sole surviving ship of Magellan's eet returned to Spain in 1522 after having circumnavigated the globe, it is said that the crew were greatly astonished that their calendar and that of the Spaniards did not correspond.

They landed, according to their own reckoning, on September 6, but were told it was September 7. At first they thought they had made a mistake, and some time elapsed before they realized that they had lost a day by going around the world with the sun. Had they traveled toward the east, they would have gained a day, and would have recorded the same date as September 8.

> "My pilot is dead of scurvy: may
> I ask the longitude, time and day?"
> The first two given and compared;
> The third,—the commandante stared!
>
> "The *first* of June? I make it second,"
> Said the stranger, "Then you've wrongly reckoned!"
>
> —Bret Harte, in The Lost Galleon.

The explanation of this phenomenon is simple. In traveling westward, in the same way with the sun, one's days are lengthened as compared with the day at any fixed place. When one has traveled 15° westward, at whatever rate of speed, he finds his watch is one hour behind the time at his starting point, if he changes it according to the sun. He has thus lost an hour as compared with the time at his starting point. After he has traveled 15° farther, he will set his watch back two hours and thus record a loss of two hours. And so it continues throughout the twenty-four belts of 15° each, losing one hour in each belt; by the time he arrives at his starting point again, he has set his hour hand back twenty-four hours and has lost a day.

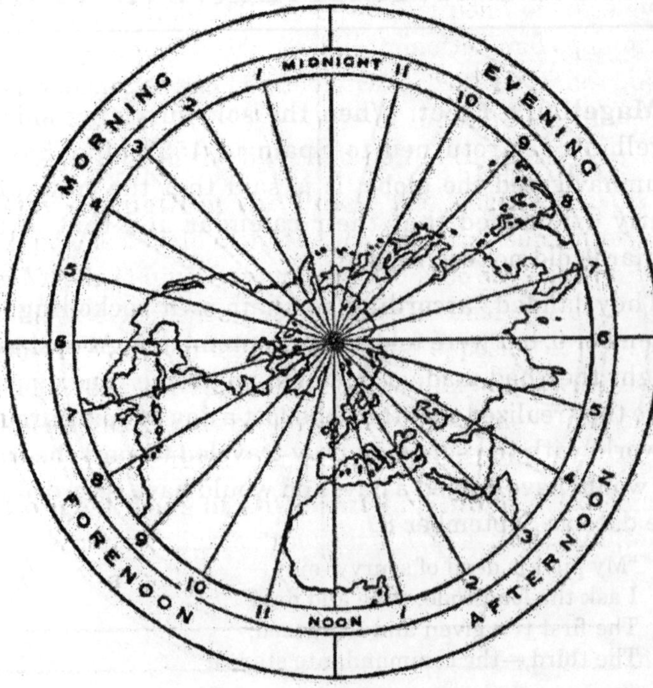

Fig. 30

Westward Travel—Days are Lengthened. To make this clearer, let us suppose a traveler starts from London Monday noon, January 1st, traveling westward 15° each day. On Tuesday, when he finds he is 15° west of London, he sets his watch back an hour. It is then noon by the sun where he is. He says, "I left Monday noon, it is now Tuesday noon; therefore I have been out one day." The tower clock at London and his chronometer set with it, however, indicate a different view. They say it is Tuesday, 1 o'clock, p.m., and he has been out a day and an hour. The next day the process is repeated. The traveler having covered another space of 15. westward, sets his watch back a second hour and says, "It is Wednesday noon and I have been out just two days." The London clock, however, says Wednesday, 2 o'clock, p.m.—two days and two hours since he left. The third day this occurs again, the traveler losing a third hour; and what to him seems three days, Monday noon to Thursday noon, is in reality by London time three days and three hours. Each of his days is really a little more than twenty-four hours long, for he is going with the sun. By the time he arrives at London again he finds what to him was twenty-four days is, in reality, twenty-five days, for he has set his watch back an hour each day for twenty-four days, or an entire day. To

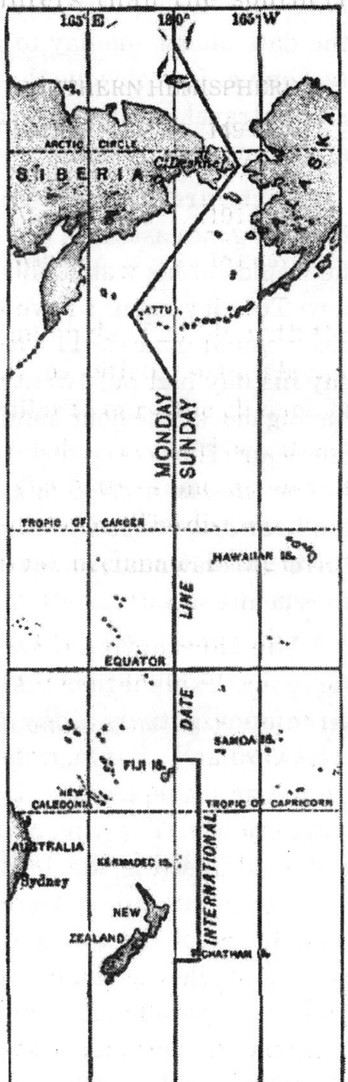

Fig. 31

have his calendar correct, he must omit a day, that is, move the date ahead one day to make up the date lost from his reckoning. It is obvious that this will be true whatever the rate of travel, and the day can be omitted from his calendar anywhere in the journey and the error corrected.

Eastward Travel—Days are Shortened. Had our traveler gone eastward, when he had covered 15° of longitude he would set his watch ahead one hour and then say, "It is now Tuesday noon. I have been out one day." The London clock would indicate 11 o'clock, a.m., of Tuesday, and thus say his day had but twenty-three hours in it, the traveler having moved the hour hand ahead one space. He has gained one hour. The second day he would gain another hour, and by the time he arrived at London again, he would have set his hour hand ahead twenty-four hours or one full day. To correct his calendar, somewhere on his voyage he would have to repeat a day.

The International Date Line. It is obvious from the foregoing explanation that somewhere and sometime in circumnavigation, a day must be omitted in traveling westward and a day repeated in traveling eastward. Where and when the change is made is a mere matter of convenience. The theoretical location of the date line commonly used is the 180th meridian. This line where a traveler's calendar needs changing varies as do the boundaries of the standard time belts and for the same reason. While the change could be made at any particular point on a parallel, it would make a serious inconvenience were the change made in some places. Imagine, for example, the 90th meridian, west of Greenwich, to be the line used. When it was Sunday in Chicago, New York, and other eastern points, it would be Monday in St. Paul, Kansas City, and western points. A traveler leaving Minneapolis on Sunday night would arrive in Chicago on Sunday morning and thus have two Sundays on successive days. Our national holidays and elections would then occur on different days in different parts of the country. To reduce to the minimum such inconveniences as necessarily attend changing one's calendar, the change is made where there is a relatively small amount of travel, away out in the Pacific Ocean. Going westward across this line one must set his calendar ahead a day; going eastward, back a day.

As shown in Figures 31 and 32, this line begins on the 180th meridian far to the north, sweeps to the eastward around Cape Deshnef, Russia, then westward beyond the 180th meridian seven degrees that the Aleutian islands may be to the east of it and have the same day as continental United States; then the line extends to the 180th meridian which it follows southward, sweeping somewhat eastward to give the Fiji and Chatham islands the same day as Australia and New Zealand. The following is a letter, by C. B. T. Moore, commander, U. S. N., Governor of Tutuila, relative to the accuracy of the map in this book:

Pago-Pago, Samoa, December 1, 1906.

Dear Sir:—The map of your Mathematical Geography is correct in placing Samoa to the east of the international date line. The older geographies were also right in placing these islands west of the international date line, because they used to keep the same date as Australia and New Zealand, which are west of the international date line.

The reason for this mistake is that when the London Missionary Society sent its missionaries to Samoa they were not acquainted with the trick of changing the date at the 180th meridian, and so carried into Samoa, which was east of the date line, the date they brought with them, which was, of course, one day ahead.

This false date was in force at the time of my first visit to Samoa, in 1889. While I have no record to show when the date was corrected, I believe that it was corrected at the time of the annexation of the Samoan Islands by the United States and by Germany. The date in Samoa is, therefore, the same date as in the United States, and is one day behind what it is in Australia and New Zealand;

Example: To-day is the 2d day of December in Auckland, and the 1st day of December in Tutuila.

Very respectfully,

C. B. T. Moore,
Commander, U. S. Navy,
Governor.

Mr. Willis E. Johnson,
Vice President Northern Normal and Industrial School,
Aberdeen, South Dakota.

"It is fortunate that the 180th meridian falls where it does. From Siberia to the Antarctic continent this imaginary line traverses nothing but water. The only land which it passes at all near is one of the archipelagoes of the south Pacific; and there it divides but a handful of volcanoes and coral reefs from the main group. These islands are even more unimportant to the world than insignificant in size. Those who tenant them are few, and those who are bound to these few still fewer There, though time ows ceaselessly on, occurs that unnatural yet unavoidable jump of twenty-four hours; and no one is there to be startled by the fact,—no one to be perplexed in trying to reconcile the two incongruities, continuous time and discontinuous day. There is nothing but the ocean, and that is tenantless Most fortunate was it, indeed, that opposite the spot where man was most destined to think there should have been placed so little to think about."*

Where Days Begin. When it is 11.30 o'clock, p.m., on Saturday at Denver, it is 1.30 o'clock, a.m., Sunday, at New York, It is thus evident that parts of two days exist at the same time on the earth. Were one to travel around the earth with the sun and as rapidly it would be perpetually noon. When he has gone around once, one day has passed. Where did that day begin? Or, suppose we wished to be the first on earth to hail the new year, where could we go to do so? The

*From *Chosön*, by Percival Lowell.

midnight line, just opposite the sun, is constantly bringing a new day somewhere. Midnight ushers in the new year at Chicago. Previous to this it was begun at New York. Still east of this, New Year's Day began some time before. If we keep going around eastward we must surely come to some place where New Year's Day was first counted, or we shall get entirely around to New York and find that the New Year's Day began the day before, and this midnight would commence it again. As previously stated, the date line commonly accepted nearly coincides with the 180th meridian. Here it is that New Year's Day first dawns and each new day begins.

The Total Duration of a Day. While a day at any particular place is twenty-four hours long, each day lasts on earth at least forty-eight hours. Any given day, say Christmas, is first counted as that day just west of the date line. An hour later Christmas begins 15° west of that line, two hours later it begins 30° west of it, and so on around the globe. The people just west of the date line who first hailed Christmas have enjoyed twelve hours of it when it begins in England, eighteen hours of it when it begins in central United States, and twenty-four hours of it, or the whole day, when it begins in western Alaska, just east of the date line. Christmas, then, has existed twenty-four hours on the globe, but having just begun in western Alaska, it will tarry twenty-four hours longer among mankind, making forty-eight hours that the day blesses the earth.

If the date line followed the meridian 180° without any variation, the total duration of a day would be exactly forty-eight hours as just explained. But that line is quite irregular, as previously described and as shown on the map. Because of this irregularity of the date line the same day lasts somewhere on earth over *forty-nine hours*. Suppose we start at Cape Deshnef, Siberia, longitude 169° West, a moment after midnight of the 3d of July. The 4th of July has begun, and, as midnight sweeps around westward, successive places see the beginning of this day. When it is the 4th in London it has been the 4th at Cape Deshnef twelve hours and forty-four minutes. When the glorious day arrives at New York, it has been seventeen hours and forty-four minutes since it began at Cape Deshnef. When it reaches our most western point on this continent, Attu Island, 173° E., it has been twenty-five hours and twelve minutes since it began at Cape Deshnef. Since it will last twenty-four hours at Attu Island, forty-nine hours and twelve minutes will have elapsed since the beginning of the day until the moment when all places on earth cease to count it that day.

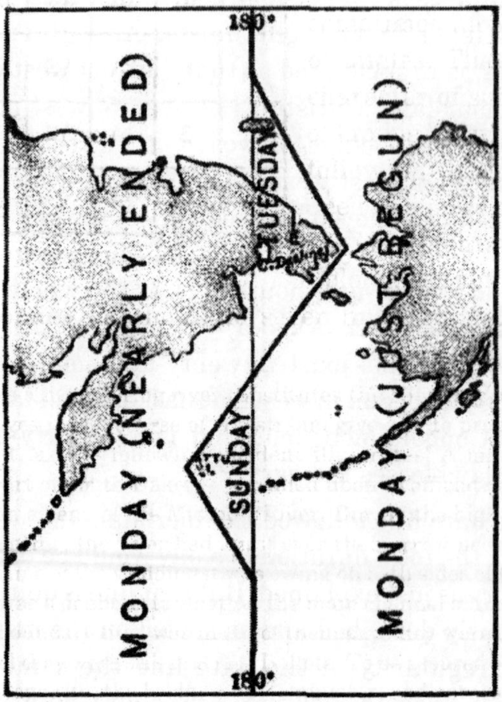

Fig. 32

When Three Days Coëxist. Portions of three days exist at the same time between 11.30 o'clock, a.m., and 12.30 o'clock, p.m., London time. When it is Monday noon at London, Tuesday has begun at Cape Deshnef, but Monday morning has not yet dawned at Attu Island; nearly half an hour of Sunday still remains there.

Confusion of Travelers. Many stories are told of the confusion to travelers who pass from places reckoning one day across this line, to places having a different day. "If it is such a deadly sin to work on Sunday, one or the other of Mr. A and Mr. B coming one from the east, the other from the west of the 180th meridian, must, if he continues his daily vocations, be in a bad way. Some of our people in the Fiji are in this unenviable position, as the line 180° passes through Loma-Loma. I went from Fiji to Tonga in Her Majesty's ship *Nymph* and arrived at our destination on Sunday, according to our reckoning from Fiji, but on Saturday, according to the proper computation west from Greenwich. We, however, found the natives all keeping Sunday. On my asking the missionaries about it they told me that the missionaries to that group and Samoa having come from the westward, had determined to observe their Sabbath day, as usual, so as not to subject the natives to any puzzle, and agreed to put the dividing line farther o", between them and Hawaii, somewhere in the broad ocean where no metaphysical natives or `intelligent' Zulus could cross-question them."*

"A party of missionaries bound from China, sailing west, and nearing the line without their knowledge, on Saturday posted a notice in the cabin announcing that 'To-morrow being Sunday there will be services in this cabin at 10 a.m.' The following morning at 9, the captain tacked up a notice

*Mr. E. L. Layard, at the British Consulate, Noumea, New Caledonia, as quoted in a pamphlet on the International Date Line by Henry Collins.

declaring that 'This being Monday there will be no services in this cabin this morning.'"

It should be remembered that this line, called "international," has not been adopted by all nations as a hard and fast line, making it absolutely necessary to change the date the moment it is crossed. A ship sailing, say, from Honolulu, which has the same day as North America and Europe, to Manila or Hongkong, having a day later, may make the change in date at any time between these distant points; and since several days elapse in the passage, the change is usually made so as to have neither two Sundays in one week nor a week without a Sunday. Just as the traveler in the United States going from a place having one time standard to a place having a different one would find it necessary to change his watch but could make the change at any time, so one passing from a place having one day to one reckoning another, could suit his convenience as to the precise spot where he make the change. This statement needs only the modification that as all events on a ship must be regulated by a common timepiece, changed according to longitude, so the community on board in order to adjust to a common calendar must accept the change when made by the captain.

Origin and Change of Date Line. The origin of this line is of considerable interest. The day adopted in any region depended upon the direction from which the people came who settled the country. For example, people who went to Australia, Hongkong, and other English possessions in the Orient traveled around Africa or across the Mediterranean. They thus set their watches ahead an hour for every 15° "For two centuries after the Spanish settlement the trade of Manila with the western world was carried on *via* Acapulco and Mexico" (Ency. Brit.). Thus the time which obtained in the Philippines was found by setting watches backwards an hour for every 15°, and so it came about that the calendar of the Philippines was a day earlier than that of Australia, Hongkong, etc. The date line at that time was very indefinite and irregular. In 1845 by a decree of the Bishop of Manila, who was also Governor-General, Tuesday, December 31, was stricken from the calendar; the day after Monday, December 30, was Wednesday, January 1, 1846. This cutting the year to 364 days and the week to 6 days gave the Philippines the same day as other Asiatic places, and shifted the date line to the east of that archipelago. Had this change never been made, all of the possessions of the United States would have the same day.

For some time after the acquisition of Alaska the people living there, formerly citizens of Russia, used the day later than ours, and also used the Russian or Julian calendar, twelve days later than ours. As people moved there from the United States, our system gradually was extended, but for a time both systems were in vogue. This made a"airs confusing, some keeping Sunday when others reckoned the same day as Saturday and counted it as twelve days later in the calendar, New Year's Day, Christmas, etc., coming at different times. Soon, however, the American system prevailed to the entire exclusion of the Russian, the inhabitants repeating a day, and thus having eight days in one week. While the Russians in their churches in Alaska are celebrating the Holy Mass on our Sunday, their brethren in Siberia, not far away, and in other parts of Russia, are busy with Monday's duties.

Date Line East of Fiji Islands. Fiji, No XIV, 1879: An ordinance enacted by the governor of the colony of Fiji, with the advice and consent of the legislative council thereof, to provide for a universal day throughout the colony.

Whereas, according to the ordinary rule of noting time, any given time would in that part of the colony lying to the east of the meridian of 180° from Greenwich be noted as of a day of the week and month different from the day by which the same time would be noted in the part of the colony lying to the west of such meridian; and

Whereas, by custom the ordinary rule has been set aside and time has been noted throughout the colony as though the whole were situated to the west of such meridian; and

Whereas, in order to preclude uncertainty for the future it is expedient that the above custom should be legalized; therefore

Be it enacted by the governor, with the advice and consent of the legislative council, as follows:

Time in this colony shall be noted as if the whole colony were situated to the west of the meridian of 180° from Greenwich.

(*Exempli gratia*—To-day, which according to the ordinary rule for noting time is on the island of Ovalau the 5th day of June, and on the island of Vanua-Balevu the 4th day of June, would by this ordinance be deemed as the 5th day of June, 1879, in the whole colony.)

Problem. Assuming it was 5 a.m., Sunday, May 1, 1898, when the naval battle of Manila began, what time was it in Milwaukee, the city using standard time and Dewey using the local time of 120° east?

CHAPTER-VI

THE EARTH'S REVOLUTION

PROOFS OF REVOLUTION

For at least 2400 years the theory of the revolution of the earth around the sun has been advocated, but only in modern times has the fact been demonstrated beyond successful contradiction. The proofs rest upon three sets of astronomical observations, all of which are of a delicate and abstruse character, although the underlying principles are easily understood.

Aberration of Light. When rain is falling on a calm day the drops will strike the top of one's head if he is standing still in the rain; but if one moves, the direction of the drops will seem to have changed, striking one in the face more and more as the speed is increased (Fig. 33). Now light rays from the sun, a star, or other heavenly body, strike the earth somewhat slantingly, because the earth is moving around the sun at the rate of over a thousand miles per minute. Because of this fact the astronomer must tip his telescope slightly to the east of a star in order to see it when the earth is in one side of its orbit, and to the west of it when in the opposite side of the orbit. The necessity of this tipping of the telescope will be apparent if we imagine the rays passing through the telescope are like raindrops falling through a tube. If the tube is carried forward swiftly enough the drops will strike the sides of the tube, and in order that they may pass directly through it, the tube must be tilted forward somewhat, the amount varying with (*a*) the rate of its onward motion, and (*b*) the rate at which the raindrops are falling.

Since the telescope must at one time be tilted one way to see a star and at another season tilted an equal amount in the opposite direction, each star thus seems to move about in a tiny orbit, varying from a circle to a straight line, depending upon the position of the star, but in every case the major axis is 41'', or twice the greatest angle at which the telescope must be tilted forward.

Fig. 33

Each of the millions of stars has its own apparent aberrational orbit, no two being exactly alike in form, unless the two chance to be exactly the same distance from the plane of the earth's orbit. Assuming that the earth revolves around the sun, the precise form of this aberrational orbit of any star can be calculated, and observation invariably confirms the calculation. Rational minds cannot conceive that the millions of stars, at varying distances, can all actually have these peculiar annual motions, six months toward the earth and six months from it, in addition to the other motions which many of them (and probably all of them, see p. 265) have. The discovery and explanation of these facts in 1727 by James Bradley (see p. 277), the English Astronomer Royal, forever put at rest all disputes as to the revolution of the earth.

Motion in the Line of Sight. If you have stood near by when a swiftly moving train passed with its bell ringing, you may have noticed a sudden change in the tone of the bell; it rings a lower note immediately upon passing. The pitch of a note depends upon the rate at which the sound waves strike the ear; the more rapid they are, the higher is the pitch. Imagine a boy throwing chips* into a river at a uniform rate while walking down stream toward a bridge and then while walking upstream away from the bridge. The chips will be closer together as they pass under the bridge when the boy is walking toward it than when he is walking away from it. In a similar way the sound waves from the bell of the rapidly approaching locomotive accumulate upon the ear of the listener, and the pitch is higher than it would be if the train were stationary, and after the train passes the sound waves will be farther apart, as observed by the same person, who will hear a lower note in consequence.

Color varies with Rate of Vibration. Now in a precisely similar manner the colors in a ray of light vary in the rate of

*This illustration is adapted from Todd's New Astronomy, p. 432.

vibration. The violet is the most rapid,* indigo about one tenth part slower, blue slightly slower still, then green, yellow, orange, and red. The spectroscope is an astronomical instrument which spreads out the line of light from a celestial body into a band and breaks it up into its several colors. If a ringing bell rapidly approaches us, or if we approach it, the tone of the bell sounds higher than if it recedes from us or if we recede from it. If we rapidly approach a star, or a star approaches us, its color shifts toward the violet end of the spectroscope; and if we rapidly recede from it, or it recedes from us, its color shifts toward the red end. Now year after year the thousands of stars in the vicinity of the plane of the earth's pathway show in the spectroscope this change toward violet at one season and toward red at the opposite season. The farther from the plane of the earth's orbit a star is located, the less is this annual change in color, since the earth neither approaches nor recedes from stars toward the poles. Either the stars near the plane of the earth's orbit move rapidly toward the earth at one season, gradually stop, and six months later as rapidly recede, and stars away from this plane approach and recede at rates diminishing exactly in proportion to their distance from this plane, *or the earth itself swiftly moves about the sun.*

*The rate of vibration per second for each of the colors in a ray of light is as follows:

Violet	756.0×10^{12}	Yellow	508.8×10^{12}
Indigo	698.8×10^{12}	Orange	457.1×10^{12}
Blue	617.1×10^{12}	Red	393.6×10^{12}
Green	569.2×10^{12}		

Thus the violet color has 756.0 millions of millions of vibrations each second; indigo, 698.8 millions of millions, etc.

Proof of the Rotation of the Earth. The same set of facts and reasoning applies to the rotation of the earth. In the evening a star in the east shows a color approaching the violet side of the spectroscope, and this gradually shifts toward the red during the night as the star is seen higher in the sky, then nearly overhead, then in the west. Now either the star swiftly approaches the earth early in the evening, then gradually pauses, and at midnight begins to go away from the earth faster and faster as it approaches the western horizon, or the earth rotates on its axis, toward a star seen in the east, neither toward nor from it when nearly overhead, and away from it when seen near the west. Since the same star rises at different hours throughout the year it would have to y back and forth toward and from the earth, two trips every day, varying its periods according to the time of its rising and setting. Besides this, when a star is rising at Calcutta it shows the violet tendency to observers there (Calcutta is rotating toward the star when the star is rising), and at the same moment the same star is setting at New Orleans and thus shows a shift toward the red to observers there. Now the distant star cannot possibly be actually rapidly approaching Calcutta and at the same time be as rapidly receding from New Orleans. The spectroscope, that wonderful instrument which has multiplied astronomical knowledge during the last half century, demonstrates, with mathematical certainty, the rotation of the earth, and multiplies millionfold the certainty of the earth's revolution.

Actual Motions of Stars. Before leaving this topic we should notice that other changes in the colors of stars show that some are actually approaching the earth at a uniform rate, and some are receding from it. Careful observations at long intervals show other changes in the positions of stars. The latter motion of a star is called its *proper motion* to distinguish it from the apparent motion it has in common with other stars due to the motions of the earth. The spectroscope also assists in the demonstration that the sun with the earth and the rest of the planets and their attendant satellites is moving rapidly toward the constellation Hercules.

Elements of Orbit Determined by the Spectroscope. As an instance of the use of the spectroscope in determining motions of celestial bodies, we may cite the recent calculations of Professor Kästner, Director of the Bonn Observatory. Extending from June 24, 1904, to January 15, 1905, he made careful observations and photographs of the spectrographic lines shown by Arcturus. He then made calculations based upon a microscopic examination of the photographic plates, and was able to determine (*a*) the size of the earth's orbit, (*b*) its form, (*c*) the rate of the earth's motion, and (*d*) the rate at which the solar system and Arcturus are approaching each other (10, 849 miles per hour, though not in a direct line).

The Parallax of Stars. Since the days of Copernicus (1473–1543) the theory of the revolution of the earth around the sun has been very generally accepted. Tycho Brahe (1546– 1601), however, and some other astronomers, rejected this theory because they argued that if the earth had a motion across the great distance claimed for its orbit, stars would change their positions in relation to the earth, and they could detect no such change. Little did they realize the tremendous distances of the stars. It was not until 1838 that an astronomer succeeded in getting the orbital or heliocentric parallax of a star. The German astronomer Bessel then discovered that the faint star 61 Cygni is annually displaced to the extent of 0.40''. Since then about forty stars have been found to have measurable parallaxes, thus multiplying the proofs of the motion of the earth around the sun.

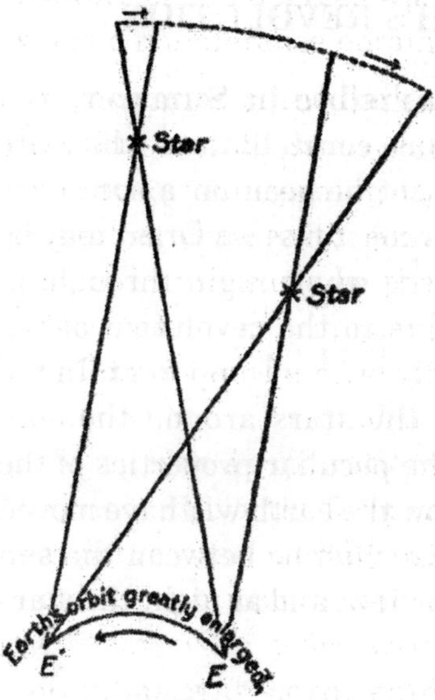

Fig. 34

Displacement of a Star Varies with its Distance. Figure 34 shows that the amount of the displacement of a star in the background of the heavens owing to a change in the position of the earth, varies with the distance of the star. The nearer the star, the greater the displacement; in every instance, however, this apparent shifting of a star is exceedingly minute, owing to the great distance (see pp. 44, 247) of the very nearest of the stars.

Since students often confuse the apparent orbit of a star described under aberration of light with that due to the parallax, we may make the following comparisons:

Aberrational Orbit	Parallactic Orbit
1. The earth's rapid motion causes the rays of light to slant (apparently) into the telesc ope so that, as the earth changes its direction in going around the sun, the star seems to shift slightly about.	1. As the earth moves about in its orbit the stars seem to move about upon the background of the celestial sphere.
2. This orbit has the same maximum width for all stars, however near or distant.	2. This orbit varies in width with the distance of the star; the nearer the star, the greater the width.

EFFECTS OF EARTH'S REVOLUTION

Winter Constellations Invisible in Summer. You have doubtless observed that some constellations which are visible on a winter's night cannot be seen on a summer's night. In January, the beautiful constellation Orion may be seen early in the evening and the whole night through; in July, not at all. That this is due to the revolution of the earth around the sun may readily be made apparent. In the daytime we cannot easily see the stars around the sun, because of its great light and the peculiar properties of the atmosphere; six months from now the earth will have moved halfway around the sun, and we shall be between the sun and the stars he now hides from view, and at night the stars now invisible will be visible.

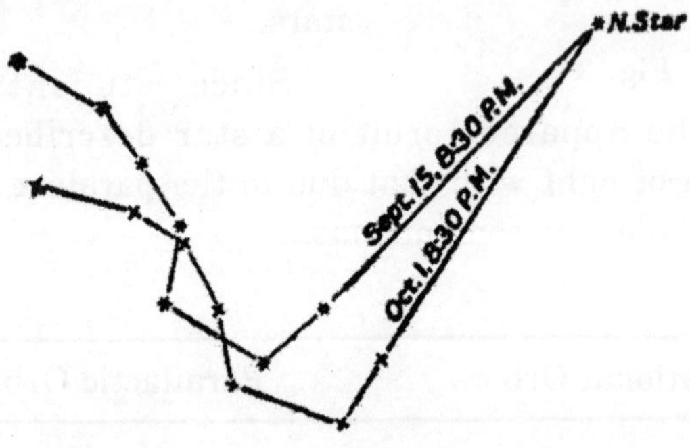

Fig. 35

If you have made a record of the observations suggested in Chapter I, you will now find that Exhibit I (Fig. 35), shows that the Big Dipper and other star groups have slightly changed their relative positions for the same time of night, making a little more than one complete rotation during each twenty-four hours. In other words, the stars have been gaining a little on the sun in the apparent daily swing of the celestial sphere around the earth.

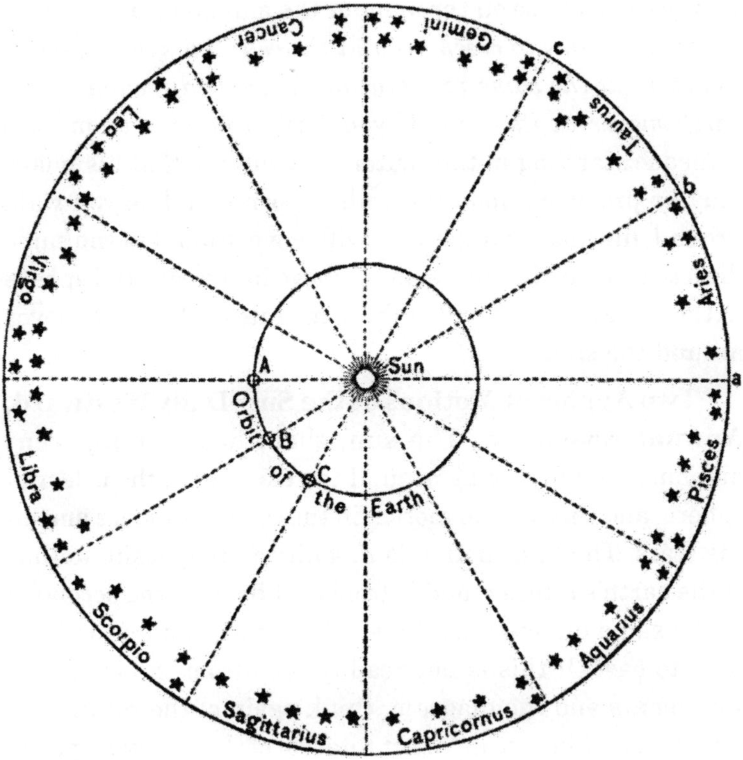

Fig. 36

The reasons for this may be understood from a careful study of Figure 36. The outer circle, which should be indefinitely great, represents the celestial sphere; the inner ellipse, the path of the earth around the sun. Now the sun does not seem to be, as it really is, relatively near the earth, but is projected into the celestial sphere among the stars. When the earth is at point *A* the sun is seen among the stars at a; when the earth has moved to *B* the sun seems to have moved to b, and so on throughout the annual orbit. *The sun, therefore, seems to creep around the celestial sphere among the stars at the same rate and in the same direction as the earth moves in its orbit.* If you walk around a room with someone standing in the center, you will see that his image may be projected upon the wall opposite, and as you walk around, his image on the wall will move around in the same direction. Thus the sun seems to move in the celestial sphere in the same direction and at the same rate as the earth moves around the sun.

Two Apparent Motions of the Sun: Daily Westward, Annual Eastward. The sun, then, has two apparent motions,—a daily swing around the earth with the celestial sphere, and this annual motion in the celestial sphere among the stars. The first motion is in a direction opposite to that of the earth's rotation and is from east to west, the second is in the same direction as the earth's revolution and is from west to east. If this is not readily seen from the foregoing statements and the diagram, think again of the rotation of the earth making an apparent rotation of the celestial sphere in the opposite direction, the reasons why the sun and moon seem to rise in the east and set in the west; then think of the motion of the earth around the sun by which the sun is projected among certain stars and then among other stars, seeming to creep among them from west to east.

After seeing this clearly, think of yourself as facing the rising sun and a star which is also rising. Now imagine the earth to have rotated once, a day to have elapsed, and the earth to have gone a day's journey in its orbit in the direction corresponding to upward. The sun would not then be on the horizon, but, the earth having moved "upward," it would be somewhat below the horizon. The same star, however, would be on the horizon, for the earth does not change its position in relation to the stars. After another rotation the earth would be, relative to the stars viewed in that direction, higher up in its orbit and the sun farther below the horizon when the star was just rising. In three months when the star rose the sun would be nearly beneath one's feet, or it would be midnight; in six months we should be on the other side of the sun, and it would be setting when the star was rising; in nine months the earth would have covered the "downward" quadrant of its journey around the sun, and the star would rise at noon; twelve months later the sun and star would rise together again. If the sun and a star set together one evening, on the next evening the star would set a little before the sun, the next night earlier still.

Since the sun passes around its orbit, 360°, in a year, 365 days, it passes over a space of nearly one degree each day. The diameter of the sun as seen from the earth covers about half a degree of the celestial sphere. During one rotation of the earth, then, the sun creeps eastward among the stars about twice its own width. A star rising with the sun will gain on the sun nearly $\frac{1}{360}$ of a day during each rotation, or a little less than four minutes. The sun sets nearly four minutes later than the star with which it set the day before.

Sidereal Day. Solar Day. The time from star-rise to star-rise, or an exact rotation of the earth, is called a *sidereal day*. Its exact length is 23 h. 56 m. 4.09 s. The time between two successive passages of the sun over a given meridian, or from noon by the sun until the next noon by the sun, is called a *solar day*.* Its length varies somewhat, for reasons to be

*A solar day is sometimes defined as the interval from sunrise to sunrise again. This is true only at the equator. The length of the solar day corresponding to February 12, May 15, July 27, or November 3, is almost exactly twenty-four hours. The time intervening between sunrise

explained later, but averages twenty-four hours. When we say "day," if it is not otherwise qualified, we usually mean an average solar day divided into twenty-four hours, from midnight to midnight. The term "hour," too, when not otherwise qualified, refers to one twenty-fourth of a mean solar day.

Fig. 37

Causes of Apparent Motions of the Sun. The apparent motions of the sun are due to the real motions of the earth. If the earth moved slowly around the sun, the sun would appear to move slowly among the stars. Just as we know the direction and rate of the earth's rotation by observing the direction and rate of the apparent rotation of the celestial sphere, we know the direction and rate of the earth's revolution by observing the direction and rate of the sun's apparent annual motion.

The Ecliptic. The path which the center of the sun seems to trace around the celestial sphere in its annual orbit

and sunrise again varies greatly with the latitude and season. On the dates named a solar day at the pole is twenty-four hours long, as it is everywhere else on earth. The time from sunrise to sunrise again, however, is almost six months at either pole.

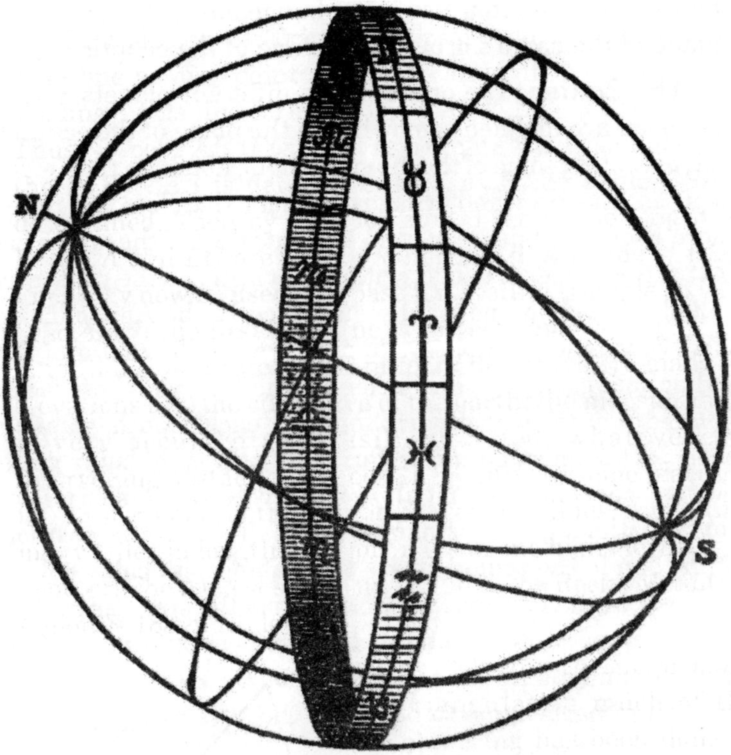

Fig. 38. Celestial sphere, showing zodiac

is called the *ecliptic.** The line traced by the center of the earth in its revolution about the sun is its orbit. Since the sun's apparent annual revolution around the sky is due to the earth's actual motion about the sun, the path of the sun, the ecliptic, must lie in the same plane with the earth's orbit. The earth's equator and parallels, if extended, would coincide with the celestial equator and parallels; similarly, the earth's orbit, if expanded in the same plane, would coincide with

*So called because eclipses can occur only when the moon crosses the plane of the ecliptic.

the ecliptic. We often use interchangeably the expressions "plane of the earth's orbit" and "plane of the ecliptic."

The Zodiac. The orbits of the different planets and of the moon are inclined somewhat to the plane of the ecliptic, but, excepting some of the minor planets, not more than eight degrees. The moon and principal planets, therefore, are never more than eight degrees from the pathway of the sun. This belt sixteen degrees wide, with the ecliptic as the center, is called the zodiac (more fully discussed in the Appendix, p. 291). Since the sun appears to pass around the center of the zodiac once each year, the ancients, who observed these facts, divided it into twelve parts, one for each month, naming each part from some constellation in it. It is probably more nearly correct historically, to say that these twelve constellations got their names originally from the position of the sun in the zodiac. Libra, the Balance, probably got its name from the fact that in ancient days the sun was among the group of stars thus named about September 23, when the days and nights are equal, thus balancing. In some such way these parts came to be called the "twelve signs of the zodiac," one for each month.

The facts in this chapter concerning the apparent annual motion of the sun were well known to the ancients, possibly even more generally than they are to-day. The reason for this is because there were few calendars and almanacs in the earlier days of mankind, and people had to reckon their days by noting the position of the sun. Thus, instead of saying that the date of his famous journey to Canterbury was about the middle of April, Chaucer says it was

> When Zephirus eek with his sweete breeth
> Enspired hath in every holt and heath
> The tendre croppes, and the younge sonne
> Hath in the Ram his halfe course yronne.

Even if clothed in modern English such a description

would be unintelligible to a large proportion of the students of to-day, and would need some such translation as the following.

"When the west wind of spring with its sweet breath hath inspired or given new life in every field and heath to the tender crops, and the young sun (young because it had got only half way through the sign Aries, the Ram, which marked the beginning of the new year in Chaucer's day) hath run half his course through the sign the Ram."

Obliquity of the Ecliptic. The orbit of the earth is not at right angles to the axis. If it were, the ecliptic would coincide with the celestial equator. The plane of the ecliptic and the plane of the celestial equator form an angle of nearly* 23½°.

This is called the obliquity of the ecliptic. We sometimes speak of this as the inclination of the earth's axis from a perpendicular to the plane of its orbit.

Since the plane of the ecliptic forms an angle of 23½° with the plane of the equator, the sun in its apparent annual course around in the ecliptic crosses the celestial equator twice each year, and at one season gets 23½° north of it, and at the opposite season 23½° south of it. The sun thus never gets nearer the pole of the celestial sphere than 66½°.

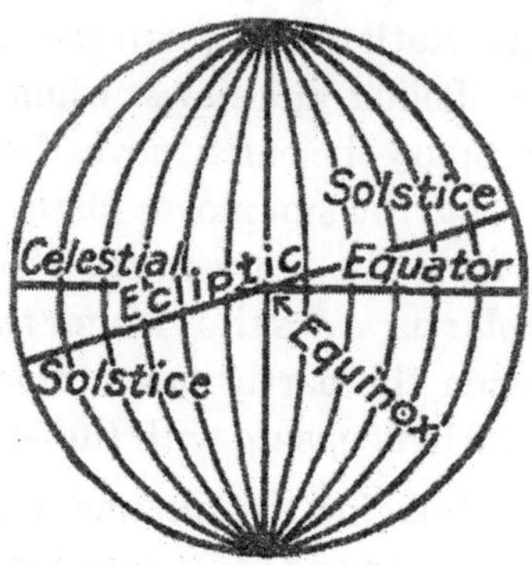

Fig. 39

*The exact amount varies slightly from year to year. The following table is taken from the Nautical Almanac, Newcomb's Calculations:

1903	1906	23° 27′ 5:45°′′
1904	1907	23° 27′ 4:98°′′
1905	1908	23° 27′ 4:51°′′

On March 21 and September 23 the sun is on the celestial equator. On June 21 and December 22 the sun is 23½° from the celestial equator.

Earth's Orbit. We have learned that the earth's orbit is an ellipse, and the sun is at a focus of it. While the eccentricity is not great, and when reduced in scale the orbit does not differ materially from a circle, the difference is sufficient to make an appreciable difference in the rate of the earth's motion in different parts of its orbit. Figure 113, p. 284, represents the orbit of the earth, greatly exaggerating the ellipticity. The point in the orbit nearest the sun is called perihelion (from *peri*, around or near, and *helios*, the sun). This point is about 91½ million miles from the sun, and the earth reaches it about December 31st. The point in the earth's orbit farthest from the sun is called aphelion (from *a*, away from, and *helios*, sun). Its distance is about 94½ million miles, and the earth reaches it about July 1st.

Varying Speed of the Earth. According to the law of gravitation, the earth moves faster in its orbit when near perihelion, and slower when near aphelion. In December and January the earth moves fastest in its orbit, and during that period the sun moves fastest in the ecliptic and falls farther behind the stars in their rotation in the celestial sphere. Solar days are thus longer then than they are in midsummer when the earth moves more slowly in its orbit and more nearly keeps up with the stars.

Imagine the sun and a star are rising together January 1st. After one exact rotation of the earth, a sidereal day, the star will be rising again, but since the earth has moved rapidly in its course around the sun, the sun is somewhat farther behind the star than it would be in summer when the earth moved more slowly around the sun. At star-rise January 3d, the sun is behind still farther, and in the course of a few weeks the sun will be several minutes behind the point where it would be if the earth's orbital motion were uniform. The sun is then said to be slow of the average sun. In July the sun creeps back less rapidly in the ecliptic, and thus a solar day is more nearly the same length as a sidereal day, and hence shorter than the average.

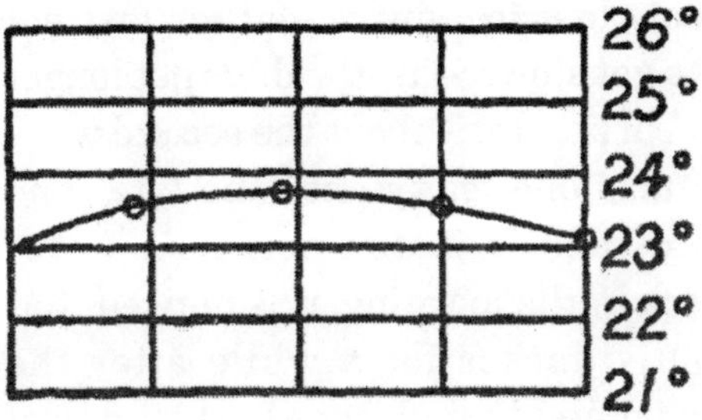

Fig. 40

Another factor modifies the foregoing statements. The daily courses of the stars swinging around with the celestial sphere are parallel and are at right angles to the axis. The sun in its annual path creeps diagonally across their courses. When farthest from the celestial equator, in June and in December, the sun's movement in the ecliptic is nearly parallel to the courses of the stars (Fig. 40); as it gets nearer the celestial equator, in March and in September, the course is more oblique. Hence in the latter part of June and of December, the sun, creeping back in the ecliptic, falls farther behind the stars and becomes slower than the average. In the latter part of March and of September the sun creeps in a more diagonal course and hence does not fall so far behind the stars in going the same distance, and thus becomes faster than the average (Fig. 41).

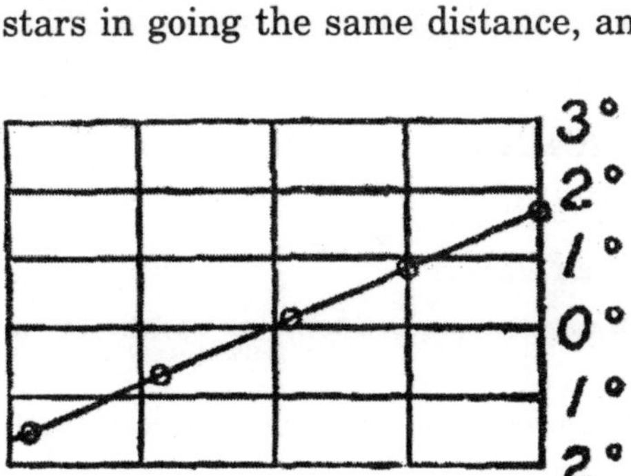

Fig. 41

Some solar days being longer than others, and the sun being sometimes slow and sometimes fast, together with standard time adoptions whereby most places have their watches set by mean solar time at some given meridian, make it unsafe to set one's watch by the sun without making many corrections.

The shortest day in the northern hemisphere is about December 22d; about that time the sun is neither fast nor slow, but it then begins to get slow. So as the days get longer the sun does not rise any earlier until about the second week of January. After Christmas one may notice the later and later time of sunsets. In schools in the northern states beginning work at 8 o'clock in the morning, it is noticed that the mornings are actually darker for a while after the Christmas holidays than before, though the shortest day of the year has passed.

Sidereal Day Shorter than Solar Day. If one wanted to set his watch by the stars, he would be obliged to remember that sidereal days are shorter than solar days; if the star observed is in a certain position at a given time of night, it will be there nearly four minutes earlier the next evening. The Greek dramatist Euripides (480–407 b.c.), in his tragedy "Rhesus," makes the Chorus say:

> Whose is the guard? Who takes my turn? The first signs are setting, and the seven Pleiades are in the sky, and the Eagle glides midway through the sky. Awake! See ye not the brilliancy of the moon? Morn, morn, indeed is approaching, and hither is one of the forewarning stars.

SUMMARY

Note carefully these propositions:

1. The earth's orbit is an ellipse.
2. The earth's orbital direction is the same as the direction of its axial motion.
3. The rate of the earth's rotation is uniform, hence sidereal days are of equal length.

4. The orbit of the earth is in nearly the same plane as that of the equator.
5. The earth's revolution around the sun makes the sun seem to creep backward among the stars from west to east, falling behind them about a degree a day. The stars seem to swing around the earth, daily gaining about four minutes upon the sun.
6. The rate of the earth's orbital motion determines the rate of the sun's apparent annual backward motion among the stars.
7. The rate of the earth's orbital motion varies, being fastest when the earth is nearest the sun or in perihelion, and slowest when farthest from the sun or in aphelion.
8. The sun's apparent annual motion, backward or eastward among the stars, is greater when in or near perihelion (December 31) than at any other time.
9. The length of solar days varies, averaging 24 hours in length. There are two reasons for this variation.

   a. Because the earth's orbital motion is not uniform, it being faster when nearer the sun, and slower when farther from it.

   b. Because when near the equinoxes the apparent annual motion of the sun in the celestial sphere is more diagonal than when near the tropics.
10. Because of these two sets of causes, solar days are more than 24 hours in length from December 25 to April 15 and from June 15 to September 1, and less than 24 hours in length from April 15 to June 15 and from September 1 to December 25.

EQUATION OF TIME

Sun Fast or Sun Slow. The relation of the apparent solar time to mean solar time is called the equation of time. As just shown, the apparent eastward motion of the sun in the ecliptic is faster than the average twice a year, and slower than the average twice a year. A fictitious sun is imagined to move at a uniform rate eastward in the celestial equator, starting with the apparent sun at the vernal equinox (see Equinox in Glossary) and completing its annual course around the celestial sphere in the same time in which the sun apparently makes its circuit of the ecliptic. While, excepting four times a year, the apparent sun is fast or slow as compared with this fictitious sun which indicates mean solar time, their difference at any moment, or the equation of time, may be accurately calculated.

The equation of time is indicated in various ways. The usual method is to indicate the time by which the apparent sun is faster than the average by a minus sign, and the time by which it is slower than the average by a plus sign. The apparent time and the equation of time thus indicated, when combined, will give the mean time. Thus, if the sun indicates noon (apparent time), and we know the equation to be 7 m. (sun fast, 7 m.), we know it is 11 h. 53 m., a.m. by mean solar time.

Any almanac shows the equation of time for any day of the year. It is indicated in a variety of ways.

a. In the World Almanac it is given under the title "Sun on Meridian." The local mean solar time of the sun's crossing a meridian is given to the nearest second. Thus Jan. 1, 1908, it is given as 12 h. 3 m. 16 s. We know from this that the apparent sun is 3 m. 16 s. slow of the average on that date.

b. In the Old Farmer's Almanac the equation of time is given in a column headed "Sun Fast," or "Sun Slow."

c. In some places the equation of time is indicated by the words, "clock ahead of sun," and "clock behind sun." Of course the student knows from this that if the clock is ahead of the sun, the sun is slower than the average, and, conversely, if the clock is behind the sun, the latter must be faster than the average.

d. Most almanacs give times of sunrise and of sunset. Now half way between sunrise and sunset it is apparent noon. Suppose the sun rises at 7.24 o'clock, a.m., and sets at 4.43 o'clock, p.m.

Half way between those times is 12.03½ o'clock, the time when the sun is on the meridian, and thus the sun is 3½ minutes slow (Jan. 1, at New York).

e. The Nautical Almanac* has the most detailed and accurate data obtainable. Table II for each month gives in the column "Equation of Time" the number of minutes and seconds to be added to or subtracted from 12 o'clock noon at Greenwich for the apparent sun time. The adjoining column gives the difference for one hour to be added when the sun is gaining, or subtracted when the sun is losing, for places east of Greenwich, and *vice versa* for places west.

Whether or not the student has access to a copy of the Nautical Almanac it may be of interest to notice the use of this table.

*Prepared annually three years in advance, by the Professor of Mathematics, United States Navy, Washington, D. C. It is sold by the Bureau of Equipment at actual cost of publication, one dollar.

AT GREENWICH MEAN NOON

Day of the week.	Day of the month.	THE SUN'S				Equation of Time to be Subtracted from Mean Time	Diff. for 1 hour	Sidereal Time, or Right Ascension of Mean Sun
		Apparent Right Ascension	Diff. for 1 Hour	Apparent Declination	Diff. for 1 Hour			
		h m s				m s		h m s
Wed.	1	18 42 9.88	11.057	S. 23 5 47.3	+11.13	3 10.29	1.200	18 38 59.60
Thur.	2	18 46 35.09	11.044	23 1 6.3	12.28	3 38.93	1.188	18 42 56.16
Frid.	3	18 50 59.99	11.030	22 55 57.7	13.42	4 7.28	1.174	18 46 52.71
Sat.	4	18 55 24.54	11.015	22 50 21.8	+14.56	4 35.27	1.158	18 50 49.27
SUN.	5	18 59 48.70	10.098	22 44 18.6	15.70	5 2.87	1.141	18 54 45.83
Mon.	6	19 4 12.45	10.979	22 37 48.2	16.82	5 30.06	1.123	18 58 42.39
Tues.	7	19 8 35.74	10.959	22 30 51.0	+17.94	5 56.80	1.104	19 2 38.94
Wed.	8	19 12 58.56	10.939	22 23 27.1	19.04	6 23.06	1.083	19 6 35.50
Thur.	9	19 17 20.85	10.918	22 15 36.8	20.14	6 48.79	1.061	19 10 32.06
Frid.	10	19 21 42.61	10.895	22 7 20.2	+21.23	7 13.99	1.038	19 14 28.62
Sat.	11	19 26 3.79	10.871	21 58 37.7	22.30	7 38.62	1.014	19 18 25.17
SUN.	12	19 30 24.39	10.846	21 49 29.5	23.37	8 2.66	0.989	19 22 21.73

Part of a page from The American Ephemeris and Nautical Almanac, Jan. 1908.

This table indicates that at 12 o'clock noon, on the meridian of Greenwich on Jan. 1, 1908, the sun is slow 3 m. 10.29 s., and is losing 1.200 s. each hour from that moment. We know it is losing, for we find that on January 2 the sun is slow 3 m. 38.93 s., and by that time its rate of loss is slightly less, being 1.188 s. each hour.

Suppose you are at Hamburg on Jan. 1, 1908, when it is noon according to standard time of Germany, one hour before Greenwich mean noon. The equation of time will be the same as at Greenwich less 1.200 s. for the hour's difference, or (3 m. 10.29 s. - 1.200 s.) 3 m. 9.09 s. If you are at New York on that date and it is noon, Eastern standard time, five hours after Greenwich noon, it is obvious that the sun is 5 × 1.200 s. or 6 s. slower than it was at Greenwich mean noon. The equation of time at New York would then be 3 m. 10.29 s. + 6 s. or 3 m. 16.29 s.

f. The Analemma graphically indicates the approximate equation of time for any day of the year, and also indicates the declination of the sun (or its distance from the celestial equator). Since our year has 365¼ days, the equation of time for a given date of one year will not be quite the same as that of the same date in a succeeding year. That for 1910 will be approximately one fourth of a day or six hours later in each day than for 1909; that is, the table for Greenwich in 1910 will be very nearly correct for Central United States in 1909. Since for the ordinary purposes of the student using this book an error of a few seconds is inappreciable, the analemma will answer for most of his calculations.

The vertical lines of the analemma represent the number of minutes the apparent sun is slow or fast as compared with the mean sun. For example, the dot representing February 25 is a little over half way between the lines representing sun slow 12 m. and 14 m. The sun is then slow about 13 m. 18 s. It will be observed that April 15, June 15, September 1, and December 25 are on the central line. The equation of time is then zero, and the sun may be said to be "on time." Persons living in the United States on the 90th meridian will see the shadow due north at 12 o'clock on those days; if west of a standard time meridian one will note the north shadow when it is past 12 o'clock, four minutes for every degree; and, if east of a standard time meridian, before 12 o'clock four minutes for each degree. Since the analemma shows how fast or slow the sun is each day, it is obvious that, knowing one's longitude, one can set his watch by the sun by reference to this diagram, or, having correct clock time, one can ascertain his longitude.

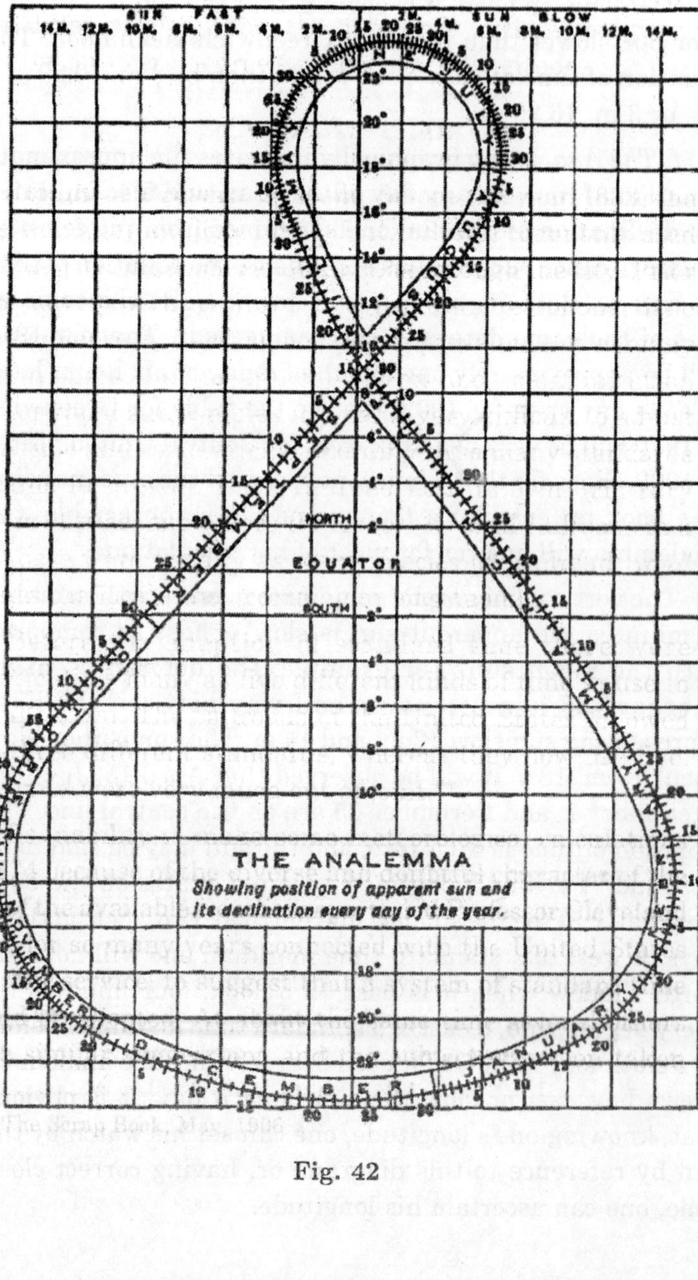
SUN FAST
SUN SLOW
THE ANALEMMA
Showing position of apparent sun and its declination every day of the year.
NORTH
EQUATOR
SOUTH

Fig. 42

USES OF THE ANALEMMA

To Ascertain Your Longitude. To do this your watch must show correct standard time. You must also have a true north-south line.

1. Carefully observe the time when the shadow is north. Ascertain from the analemma the number of minutes and seconds the sun is fast or slow.

2. If fast, add that amount to the time by your watch; if slow, subtract. This gives your mean local time.

3. Divide the minutes and seconds past or before twelve by four. This gives you the number of degrees and minutes you are from the standard time meridian. If the corrected time is before twelve, you are east of it; if after, you are west of it.

4. Subtract (or add) the number of degrees you are east (or west) of the standard time meridian, and this is your longitude.

For example, say the date is October 5th. 1. Your watch says 12 h. 10 m. 30 s., p.m., when the shadow is north. The analemma shows the sun to be 11 m. 30 s. fast. 2. The sun being fast, you add these and get 12.22 o'clock, p.m. This is the mean local time of your place. 3. Dividing the minutes past twelve by four, you get 5 m. 30 s. This is the number of degrees and minutes you are west from the standard meridian. If you live in the Central standard time belt of the United States, your longitude is 90° plus 5. 300, or 95° 30′. If you are in the Eastern time belt, it is 75° plus 5° 30′. If you are in Spain, it is 0° plus 5° 30′, and so on.

To Set Your Watch. To do this you must know your longitude and have a true north-south line.

1. Find the difference between your longitude and that of the standard time meridian in accordance with which you

wish to set your watch. In Eastern United States the standard time meridian is the 75th, in Central United States the 90th, etc.

2. Multiply the number of degrees and seconds of difference by four. This gives you the number of minutes and seconds your time is faster or slower than local time. If you are east of the standard meridian, your watch must be set slower than local time; if west, faster.

3. From the analemma observe the position of the sun whether fast or slow and how much. If fast, subtract that time from the time obtained in step two; if slow, add. This gives you the time before or after twelve when the shadow will be north; before twelve if you are east of the standard time meridian, after twelve if you are west.

4. Carefully set your watch at the time indicated in step three when the sun's shadow crosses the north-south line.

For example, suppose your longitude is 87° 37′ W. (Chicago). 1. The difference between your longitude and your standard time meridian, 90°, is 2° 23′. 2° Multiplying this difference by four we get 9° 32′, the minutes and seconds your time is slower than the sun's average time. That is, the sun on the average casts a north shadow at 11 h. 50 m. 28 s. at your longitude. 3° From the analemma we see the sun is 14 m. 15 s. slow on February 6. The time being slow, we add this to 11 h. 50 m. 28 s. and get 12 h. 4 m. 43 s., or 4 m. 43 s. past twelve when the shadow will be north. 4. Just before the shadow is north get your watch ready, and the moment the shadow is north set it 4 m. 43 s. past twelve.

To Strike a North-South Line. To do this you must know your longitude and have correct time.

Steps 1, 2, and 3 are exactly as in the foregoing explanation how to set your watch by the sun. At the time you obtain in step 3 you know the shadow is north; then draw the line of the shadow, or, if out of doors, drive stakes or otherwise indicate the line of the shadow.

To Ascertain Your Latitude. This use of the analemma is reserved for later discussion.

Civil and Astronomical Days. The mean solar day of twenty-four hours reckoned from midnight is called a civil day, and among all Christian nations has the sanction of law and usage. Since astronomers work at night they reckon a day from noon. Thus the civil forenoon is dated a day ahead of the astronomical day, the afternoon being the last half of the civil day but the beginning of the astronomical day. Before the invention of clocks and watches, the sundial was the common standard for the time during each day, and this, as we have seen, is a constantly varying one. When clocks were invented it was found impossible to have them so adjusted as to gain or lose with the sun. Until 1815 a civil day in France was a day according to the actual position of the sun, and hence was a very uncertain affair.

A FEW FACTS: DO YOU UNDERSTAND THEM?

1. A day of twenty-four hours as we commonly use the term, is not one rotation of the earth. A solar day is a little more than one complete rotation and averages exactly twenty-four hours in length. This is a civil or legal day.

2. A sidereal day is the time of one rotation of the earth on its axis.

3. There are 366 rotations of the earth (sidereal days) in one year of 365 days (solar days).

4. A sundial records apparent or actual sun time, which is the same as mean sun time only four times a year.

5. A clock records mean sun time, and thus corresponds to sundial time only four times a year.

6. In many cities using standard time the shadow of the sun is never in a north-south line when the clock strikes twelve. This is true of all cities more than 4° east or west of the meridian on which their standard time is based.

7. Any city within 4° of its standard time meridian will have north-south shadow lines at twelve o'clock no more than four times a year at the most. Strictly speaking, practically no city ever has a shadow exactly north-south at twelve o'clock.

CHAPTER-VII

TIME AND THE CALENDAR

"In the early days of mankind, it is not probable that there was any concern at all about dates, or seasons, or years. Herodotus is called the father of history, and his history does not contain a single date. Substantially the same may be said of Thucydides, who wrote only a little later—somewhat over 400 b.c. If Geography and Chronology are the two eyes of history, then some histories are blind of the one eye and can see but little out of the other."*

Sidereal Year. Tropical Year. As there are two kinds of days, solar and sidereal, there are two kinds of years, solar or tropical years, and sidereal years, but for very different reasons. The sidereal year is the time elapsing between the passage of the earth's center over a given point in its orbit until it crosses it again. For reasons not properly discussed here (see Precession of the Equinoxes, p. 285), the point in the orbit where the earth is when the vertical ray is on the equator shifts slightly westward so that we reach the point of the vernal equinox a second time a few minutes before a sidereal year has elapsed. The time elapsing from the sun's crossing of the celestial equator in the spring until the crossing the next spring is a *tropical*, year and is what

*R. W. Farland in Popular Astronomy for February, 1895.

we mean when we say "a year."* Since it is the tropical year that we attempt to fit into an annual calendar and which marks the year of seasons, it is well to remember its length: 365 d. 5 h. 48 m. 45.51 s. (365.2422 d.). The adjustment of the days, weeks, and months into a calendar that does not change from year to year but brings the annual holidays around in the proper seasons, has been a difficult task for the human race to accomplish. If the length of the year were an even number of days and that number was exactly divisible by twelve, seven, and four, we could easily have seven days in a week, four weeks in a month, and twelve months in a year and have no time to carry over into another year or month.

The Moon the Measurer. Among the ancients the moon was the great measurer of time, our word month comes from the word moon, and in connection with its changing phases religious feasts and celebrations were observed. Even to-day we reckon Easter and some other holy days by reference to the moon. Now the natural units of time are the solar day, the lunar month (about 29½ days), and the tropical year. But their lengths are prime to each other. For some reasons not clearly known, but believed to be in accordance with the four phases of the moon, the ancient Egyptians and Chaldeans divided the month into four weeks of seven days each. The addition of the week as a unit of time which is naturally related only to the day, made confusion worse confounded. Various devices have been used at different times to make the same date come around regularly in the same season year after year, but changes made by priests

*A third kind of year is considered in astronomy, the anomalistic year, the time occupied by the earth in traveling from perihelion to perihelion again. Its length is 365 d. 6 h. 13 m. 48.09 s. The lunar year, twelve new moons, is about eleven days shorter than the tropical year. The length of a sidereal year is 365 d. 6 h. 9 m. 8.97 s.

who were ignorant as to the astronomical data and by more ignorant kings often resulted in great confusion. The very exact length of the solar year in the possession of the ancient Egyptians seems to have been little regarded.

Early Roman Calendar. Since our calendar is the same as that worked out by the Romans, a brief sketch of their system may be helpful. The ancient Romans seem to have had ten months, the first being March. We can see that this was the case from the fact that September means seventh; October, eighth; November, ninth; and December, tenth. It was possibly during the reign of Numa that two months were added, January and February. There are about 29½ days in a lunar month, or fromone new moon to the next, so to have their months conform to the moons they were given 29 and 30 days alternately, beginning with January. This gave them twelve lunar months in a year of 354 days. It was thought unlucky to have the number even, so a day was added for luck.

This year, having but 355 days, was over ten days too short, so festivals that came in the summer season would appear ten days earlier each year, until those dedicated to Bacchus, the god of wine, came when the grapes were still green, and those of Ceres, the goddess of the harvest, before the heads of the wheat had appeared. To correct this an extra month was added, called Mercedonius, every second year. Since the length of this month was not fixed by law but was determined by the pontiffs, it gave rise to serious corruption and fraud, interfering with the collection of debts by the dropping out of certain expected dates, lengthening the terms of office of favorites, etc.

The Julian and the Augustan Calendars. In the year 46 b.c., Julius Caesar, aided by the Egyptian astronomer, Sosigenes, reformed the calendar. He decreed that beginning with January the months should have alternately 31 and 30 days, save February, to which was assigned 29 days, and every fourth year an additional day. This made a year of exactly 365¼ days. Since the true year has 365 days, 5 hours, 48 min., 45.51 sec., and the Julian year had 365 days, 6 hours, it was 11 min., 14.49 sec. too long.

	JULIAN	AUGUSTAN
Jan.	31	31
Feb.	29–30	28–29
Mar.	31	31
Apr.	30	30
May	31	31
June	30	30
July	31	31
Aug.	30	31
Sept.	31	30
Oct.	30	31
Nov.	31	30
Dec.	30	31

During the reign of Augustus another day was taken from February and added to August in order that that month, the name of which had been changed from Sextilis to August in his honor, might have as many days in it as the month Quintilis, whose name had been changed to July in honor of Julius Caesar. To prevent the three months, July, August, and September, from having 31 days each, such an arrangement being considered unlucky, Augustus ordered that one day be taken from September and added to October, one from November and added to December. Thus we find the easy plan of remembering the months having 31 days, every other one, was disarranged, and we must now count our knuckles or learn:

> "Thirty days hath September, April, June, and November. All the rest have thirty-one, save the second one alone, Which has four and twenty-four, till leap year gives it one day more."

The Gregorian Calendar. This Julian calendar, as it is called, was adopted by European countries just as they adopted other Roman customs. Its length was 365.25 days, whereas the true length of the year is 365.2422 days. While the error was only .0078 of a day, in the course of centuries this addition to the true year began to amount to days. By 1582 the difference had amounted to about 13 days, so that the time when the sun crosses the celestial equator, occurred the 11th of March. In that year Pope Gregory XIII reformed the calendar so that the March equinox might occur on March 21st, the same date as it did in the year 325 a.d., when the great Council of Nic a was held which finally decided the method of reckoning Easter. One thousand two hundred and fifty-seven years had elapsed, each being 11 min. 14 sec. too long. The error of 10 days was corrected by having the date following October 4th of that year recorded as October 15. To prevent a recurrence of the error, the Pope further decreed that thereafter the centurial years not divisible by 400 should not be counted as leap years. Thus the years 1600, 2000, 2400, etc., are leap years, but the years 1700, 1900, 2100, etc., are not leap years. This calculation reduces the error to a very low point, as according to the Gregorian calendar nearly 4000 years must elapse before the error amounts to a single day.

The Gregorian calendar was soon adopted in all Roman Catholic countries, France recording the date after December 9th as December 20th. It was adopted by Poland in 1786, and by Hungary in 1787. Protestant Germany, Denmark, and Holland adopted it in 1700 and Protestant Switzerland in 1701. The Greek Catholic countries have not yet adopted this calendar and are now thirteen days behind our dates. Non-Christian countries have calendars of their own.

In England and her colonies the change to the Gregorian system was effected in 1752 by having the date following September 2d read September 14. The change was violently opposed by some who seemed to think that changing the number assigned to a particular day modified time itself, and the members of the Government are said to have been mobbed in London by laborers who cried “give us back our eleven days.”

Old Style and New Style. Dates of events occurring before this change are usually kept as they were then written, the letters o.s. sometimes being written after the date to signify the old style of dating. To translate a date into the Gregorian or new style, one must note the century in which it occurred. For example, Columbus discovered land Oct. 12, 1492, o.s. According to the Gregorian calendar a change of 10 days was necessary in 1582. In 1500, leap year was counted by the old style but should not have been counted by the new style. Hence, in the century ending 1500, only 9 days difference had been made. So the discovery of America occurred October 12, o.s. or October 21, n.s. English

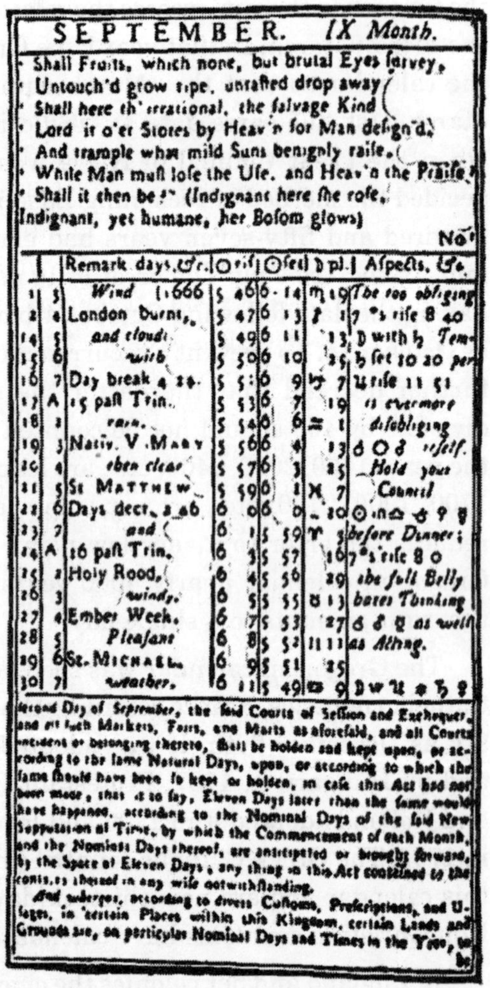

SEPTEMBER. *IX Month.*

Shall Fruits, which none, but brutal Eyes survey,
Untouch'd grow ripe, untasted drop away?
Shall here th' irrational, the salvage Kind
Lord it o'er Stores by Heav'n for Man design'd,
And trample what mild Suns benignly raise,
While Man must lose the Use, and Heav'n the Praise?
Shall it then be?" (Indignant here she rose,
Indignant, yet humane, her Bosom glows)

No

		Remark days, &c.	☉ rise	☉ sets	☽ pl.	Aspects, &c.
1	3	*Wind* 1666	5 46	6 14	♍ 19	*The too obliging*
2	4	London burnt,	5 47	6 13	♐ 1	7 *s rise 8 40
14	5	*and cloudy*	5 49	6 11	13	☽ with ♄ *Tem-*
15	6	*with*	5 50	6 10	25	♄ set 10 20 *per*
16	7	Day break 4 24.	5 51	6 9	♑ 7	♃ rise 11 51
17	A	15 past Trin.	5 53	6 7	19	*is evermore*
18	2	*rain.*	5 54	6 6	♒ 1	*disobliging*
19	3	Nativ. V. Mary	5 56	6 4	13	☌ ☉ ☿ *itself.*
20	4	*then clear*	5 57	6 3	25	*Hold your*
21	5	St Matthew	5 59	6 1	♓ 7	*Council*
22	6	Days decr. 2 26	6 0	6 0	20	☉ in ♎ ☌ ♀ ☿
23	7	*and*	6 1	5 59	♈ 3	*before Dinner;*
24	A	16 past Trin.	6 3	5 57	16	7 *s rise 8 0
25	2	Holy Rood.	6 4	5 56	29	*the full Belly*
26	3	*windy.*	6 5	5 55	♉ 13	*hates Thinking*
27	4	Ember Week.	6 7	5 53	27	☌ ♂ ☿ *as well*
28	5	*Pleasant*	6 8	5 52	♊ 11	*as Acting.*
29	6	St. Michael.	6 9	5 51	25	
30	7	*weather.*	6 11	5 49	♋ 9	☽ w ♃ ☌ ♄ ☿

second Day of *September*, the said Courts of Session and Exchequer, and all such Markets, Fairs, and Marts as aforesaid, and all Courts incident or belonging thereto, shall be holden and kept upon, or according to the same Natural Days, upon, or according to which the same should have been so kept or holden, in case this Act had not been made, that is to say, Eleven Days later than the same would have happened, according to the Nominal Days of the said New Supputation of Time, by which the Commencement of each Month, and the Nominal Days thereof, are anticipated or brought forward, by the Space of Eleven Days; any thing in this Act contained to the contrary thereof in any wise notwithstanding.

And whereas, according to divers Customs, Prescriptions, and Usages, in certain Places within this Kingdom, certain Lands and Grounds are, on particular Nominal Days and Times in the Year, to be

Fig. 43. Page from Franklin's Almanac Showing Omission of Eleven Days, 1752.

historians often write such dates October $\frac{12}{21}$, the upper date referring to old style and the lower to new style.

A historian usually follows the dates in the calendar used by his country at the time of the event. If, however, the event refers to two nations having different calendars, both dates are given. Thus, throughout Macaulay's "History of England" one sees such dates as the following: Avaux, $\frac{\text{July 27}}{\text{Aug. 6}}$, 1689. (Vol. III.) A few dates in American history prior to September, 1752, have been changed to agree with the new style. Thus Washington was born Feb. 11, 1731, o.s., but we always write it Feb. 22, 1732. The reason why all such dates are not translated into new style is because great confusion would result, and, besides, some incongruities would obtain. Thus the principal ship of Columbus was wrecked Dec. 25, 1492, and Sir Isaac Newton was born Dec. 25, 1642, and since in each case this was Christmas, it would hardly do to record them as Christmas, Jan. 3, 1493, in the former instance, or as Christmas, Jan. 4, 1643, in the latter case, as we should have to do to write them in new style.

The Beginning of the Year. With the ancient Romans the year had commenced with the March equinox, as we notice in the names of the last months, September, October, November, December, meaning 7th, 8th, 9th, 10th, which could only have those names by counting back to March as the first month. By the time of Julius Caesar the December solstice was commonly regarded as the beginning of the year, and he confirmed the change, making his new year begin January first. The later Teutonic nations for a long time continued counting the beginning of the year from March 25th. In 1563, by an edict of Charles IX, France changed the time of the beginning of the year to January first. In 1600 Scotland made the same change and England did the same in 1752 when the Gregorian system was adopted there. Dates between the first of January and the twenty-fifth of March from 1600 to 1752 are in one year in Scotland and another year in England. In Macaulay's "History of England" (Vol. III, p. 258), he gives the following reference: "Act. Parl. Scot., Mar. 19, 1689–90." The date being between January 1st and March 25th in the interval between 1600 and 1752, it was recorded as the year 1689 in England, and a year later, or 1690, in Scotland—Scotland dating the new year from January 1st, England from March 25th. This explains also why Washington's birthday was in 1731, o.s., and 1732, n.s., since English colonies used the same system of dating as the mother country.

Old Style is still used in England's Treasury Department. "The old style is still retained in the accounts of Her Majesty's Treasury. This is why the Christmas dividends are not considered due until Twelfth Day, and the midsummer dividends not till the 5th of July, and in just the same way it is not until the 5th of April that Lady Day is supposed to arrive. There is another piece of antiquity in the public accounts. In the old times, the year was held to begin on the 25th of March, and this change is also still observed in the computations over which the Chancellor of the Exchequer presides. The consequence is, that the first day of the financial year is the 5th of April, being old Lady Day, and with that day the reckonings of our annual budgets begin and end."—*London Times*,* Feb. 16, 1861.

Greek Catholic Countries Use Old Style. The Greek Catholic countries, Russia, some of the Balkan states and Greece, still employ the old Julian calendar which now, with their counting 1900 as a leap year and our not counting it

*Under the date of September 10, 1906, the same authority says that the facts above quoted obtain in England at the present time.

so, makes their dates 13 days behind ours. Dates in these countries recorded by Protestants or Roman Catholics or written for general circulation are commonly recorded in both styles by placing the Gregorian date under the Julian date. For example, the date we celebrate as our national holiday would be written by June 21 an American in Russia as $\frac{\text{June 21}}{\text{July 4}}$. The day we commemorate as the anniversary of the birth of Christ, Dec. $\frac{12}{25}$; the day they commemorate $\frac{\text{Dec. 25, 1906}}{\text{Jan 7, 1907}}$. It should be remembered that if the date is before 1900 the difference will be less than thirteen days. Steps are being taken in Russia looking to an early revision of the calendar.

Mohammedan and Jewish Calendars. The old system employed before the time of the Caesars is still used by the Mohammedans and the Jews. The year of the former is the lunar year of $354\frac{11}{30}$ days, and being about .03 of a year too short to correspond with the solar year, the same date passes through all seasons of the year in the course of 33 years. Their calendar dates from the year of the Hegira, or the flight of Mohammed, which occurred July, 622 a.d. If their year was a full solar year, their date corresponding to 1900 would be 622 years less than that number, or 1278, but being shorter in length there are more of them, and they write the date 1318, that year beginning with what to us was May 1. That is to say, what we called May 1, 1900, they called the first day of their first month, Muharram, 1318.

Chinese Calendar. The Chinese also use a lunar calendar; that is, with months based upon the phases of the moon, each month beginning with a new moon. Their months consequently have 29 and 30 days alternately. To correct the error due to so short a year, seven out of every nineteen years have thirteen months each. This still leaves the average year too short, so in every cycle of sixty years, twenty-two extra months are intercalated.

Ancient Mexican Calendar. The ancient Mexicans had a calendar of 18 months of 20 days each and five additional days, with every fourth year a leap year. Their year began with the vernal equinox.

Chaldean Calendar. Perhaps the most ancient calendar of which we have record, and the one which with modifications became the basis of the Roman calendar which we have seen was handed down through successive generations to us, was the calendar of the Chaldeans. Long before Abraham left Ur of the Chaldees (see Genesis xi, 31; Nehemiah ix, 7, etc.) that city had a royal observatory, and Chaldeans had made subdivisions of the celestial sphere and worked out the calendar upon which ours is based.

Few of us can fail to recall how hard fractions were when we first studied them, and how we avoided them in our calculations as much as possible. For exactly the same reason these ancient Chaldeans used the number 60 as their unit wherever possible, because that number being divisible by more numbers than any other less than 100, its use and the use of any six or a multiple of six avoided fractions. Thus they divided circles into 360 degrees (6 × 60), each degree into 60 minutes, and each minute into 60 seconds. They divided the zodiac into spaces of 30° each, giving us the plan of twelve months in the year. Their divisions of the day led to our 24 hours, each having 60 minutes, with 60 seconds each. They used the week of seven days, one for each of the heavenly bodies that were seen to move in the zodiac. This origin is suggested in the names of the days of the week.

DAYS OF THE WEEK

Modern English	Celestial Origin	Roman	Modern French	Ancient Saxon	Modern German
1. Sunday	Sun	Dies Solis	Dimanche	Sunnan-daeg	Sonntag
2. Monday	Moon	Dies Lunæ	Lundi	Monan-daeg	Montag
3. Tuesday	Mars	Dies Martis	Mardi	Mythical God Tiew or Tuesco Tues-daeg	Dienstag
4. Wednesday	Mercury	Dies Mercurii	Mereredi	Woden Woden's-daeg	(Mid-week) Mittwoche
5. Thursday	Jupiter	Dies Jovis	Jeudi	Thor (thunderer) Thor-daeg	Donnerstag
6. Friday	Venus	Dies Veneris	Vendredi	Friga Frigedaeg	Freitag
7. Saturday	Saturn	Dies Saturni	Samedi	Saeter-daeg	Samstag or Sonnabend

Complex Calendar Conditions in Turkey. "But it is in Turkey that the time problem becomes really complicated, very irritating to him who takes it seriously, very funny to him who enjoys a joke. To begin with, there are four years in Turkey—a Mohammedan civil year, a Mohammedan religious year, a Greek or Eastern year, and a European or Western year. Then in the year there are both lunar months depending on the changes of the moon, and months which, like ours, are certain artificial proportions of the solar year. Then the varieties, of language in Turkey still further complicate the calendars in customary use. I brought away with me a page from the diary which stood on my friend's library table, and which is customarily sold in Turkish shops to serve the purpose of a calendar; and I got from my friend the meaning of the hieroglyphics, which I record here as well as I can remember them. This page represents one day.

Numbering the compartments in it from left to right, it reads as follows:

Fig. 44

1. March, 1318 (Civil Year).
2. March, 1320 (Religious Year).
3. Thirty-one days (Civil Year).
4. Wednesday.
5. Thirty days (Religious Year).
6. 27 (March: Civil Year).
7. (March: Religious Year.)
8. March, Wednesday (Armenian).
9. April, Wednesday (French)
10. March, Wednesday (Greek)
11. Ecclesiastical Day (French R. C. Church).
12. March, Wednesday (Russian).
13. Month Day (Hebrew).
14. Month Day (Old Style).
15. Month Day (New Style).
16. Ecclesiastical Day (Armenian).
17. Ecclesiastical Day (Greek)
18. Midday, 5.35, 1902; Midday, 5.21.

"I am not quite clear in my mind now as to the meaning of the last section, but I think it is that noon according to European reckoning, is twenty-one minutes past five according to Turkish reckoning. For there is in Turkey, added to the complication of year, month, and day, a further complication as to hours. The Turks reckon, not from an artificial or conventional hour, but from sunrise, and their reckoning runs for twenty-four hours. Thus, when the sun rises at 6.30 our noon will be 5.30, Turkish time. The Turkish hours, therefore, change every day. The steamers on the Bosphorus run according to Turkish time, and one must first look in the time-table to see the hour, and then calculate from sunrise of the day what time by his European clock the boat will start. My friends in Turkey had apparently gotten used to this complicated calendar, with its variable years and months and the constantly changing hours, and took it as a matter of course."*

Modern Jewish Calendar. The modern Jewish calendar employs also a lunar year, but has alternate years lengthened by adding extra days to make up the difference between such year and the solar year. Thus one year will have 354 days, and another 22 or 23 days more. Sept. 23, 1900, according to our calendar, was the beginning of their year 5661.

Many remedies have been suggested for readjusting our calendar so that the same date shall always recur on the same day of the week. While it is interesting for the student to speculate on the problem and devise ways of meeting the difficulties, none can be suggested that does not involve so many changes from our present system that it will be impossible for a long, long time to overcome social inertia sufficiently to accomplish a reform.

If the student becomes impatient with the complexity of the problem, he may recall with profit these words of John Fiske: "It is well to simplify things as much as possible, but this world was not so put together as to save us the trouble of using our wits."

Three Christmases in One Year. "Bethlehem, the home of Christmases, is that happy Utopia of which every American child dreams—it has more than one Christmas. In fact, it has three big ones, and, strangely enough, the one falling on December 25th of our calendar is not the greatest

*The Impressions of a Careless Traveler, by Lyman Abbott.—The Outlook, Feb. 28, 1903.

of the three. It is, at least, the first. Thirteen days after the Latin has burned his Christmas incense in the sacred shrine, the Greek Church patriarch, observing that it is Christmas-time by his slower calendar, catches up the Gloria, and bows in the Grotto of the Nativity for the devout in Greece, the Balkan states, and all the Russias. After another period of twelve days the great Armenian Church of the East takes up the anthem of peace and good-will, and its patriarch visits the shrine."*

Topics for Special Reports. The gnomon. The clepsydra. Other ancient devices for reckoning time. The week. The Metonic cycle and the Golden Number. The calculation of Easter. The Roman calendar. Names of the months and days of the week. Calendar reforms. The calendar of the French Revolution. The Jewish calendars. The Turkish calendar.

*Ernest I. Lewis in Woman's Home Companion, December, 1903.

CHAPTER-VIII

SEASONS

Vertical and Slanting Rays of the Sun. He would be unobservant, indeed, who did not know from first-hand experience that the morning and evening rays of the sun do not feel so warm as those of midday, and, if living outside the torrid zone, that rays from the low winter sun in some way lack the heating power of those from the high summer sun. The reason for this difference may not be so apparent. The vertical rays are not warmer than the slanting ones, but the more nearly vertical the sun, the more heat rays are intercepted by a given surface. If you place a tub in the rain and tip it so that the rain falls in slantingly, it is obvious that less water will be caught than if the tub stood at right angles to the course of the raindrops. But before we take up in detail the e"ects of the shifting rays of the sun, let us carefully examine the conditions and causes of the shifting.

Motions of the Earth. The direction and rate of the earth's *rotation* are ascertained from the direction and rate of the apparent rotation of the celestial sphere. The direction and rate of the earth's revolution are ascertained from the apparent *revolution* of the sun among the stars of the celestial sphere. Just as any change in the rotation of the earth would produce a corresponding change in the apparent rotation of the celestial sphere, so any change in the revolution of the earth would produce a corresponding change in the apparent revolution of the sun.

Were the sun to pass among the stars at right angles to the celestial equator, passing through the celestial poles, we should know that the earth went around the sun in a path whose plane was perpendicular to the plane of the equator and was in the plane of the axis. In such an event the sun at some time during the year would shine vertically on each point on the earth's surface. Seasons would be nearly the same in one portion of the earth as in another. The sun would sometimes cast a north shadow at any given place and sometimes a south shadow. Were the sun always in the celestial equator, the ecliptic coinciding with it, we should know that the earth traveled around the sun at right angles to the axis. The vertical ray of the sun would then always be overhead at noon on the equator, and no change in season would occur. Were the plane of the earth's orbit at an angle of 45° from the equator the ecliptic would extend half way between the poles and the equator, and the sun would at one time get within 45° of the North star and six months later 45° from the South star. The vertical ray on the earth would then travel from 45° south latitude to 45° north latitude, and the torrid zone would be 90° wide.

Obliquity of the Ecliptic. But we know that the vertical ray never gets farther north or south of the equator than about 23½°, or nearer the poles than about 66½°. The plane of the ecliptic or of the earth's orbit is, then, inclined at an angle of 66½° to the axis, or at an angle of 23½° to the plane of the equator. This obliquity of the ecliptic varies slightly from year to year, as is shown on pp. 119, 286.

Equinoxes. The sun crosses the celestial equator twice a year, March 20 or 21, and September 22 or 23,* varying from year to year, the exact date for any year being easily found by referring to any almanac. These dates are called equinoxes (equinox; *æquus*, equal; *nox*, night), for the reason that the days and nights are then twelve hours long everywhere on earth. March 21 is called the vernal (spring) equinox, and September 23 is called the autumnal equinox, for reasons obvious to those who live in the northern hemisphere (see Equinox in Glossary).

Solstices. About the time when the sun reaches its most distant point from the celestial equator, for several days it seems neither to recede from it nor to approach it. The dates when the sun is at these two points are called the solstices (from *sol*, sun; and *stare*, to stand). June 21 is the summer solstice, and December 22 is the winter solstice; *vice versa* for the southern hemisphere. The same terms are also applied to the two points in the ecliptic farthest from the equator; that is, the position of the sun on those dates.

At the Equator. *March 21*. Imagine you are at the equator March 21. Bear in mind the fact that the North star (strictly speaking, the north pole of the celestial sphere) is on the northern horizon, the South star on the southern horizon, and the celestial equator extends from due east, through the zenith, to due west. It is sunrise of the vernal equinox. The sun is seen on the eastern horizon; the shadow

*The reason why the date shifts lies in the construction of our calendar, which must fit a year of 365 days, 5 h. 48 m. 45.51 s. The time of the vernal equinox in 1906 was March 21, 7:46 a.m., Eastern standard time. In 1907 it occurred 365 days, 5 h. 48 m. 45.51 s. later, or at 1:35 p.m., March 21. In 1908, being leap year, it will occur 366 days, 5 h. 48 m. 45.51 s. later, or at about 7:24 p.m., March 20. The same facts are true of the solstices; they occur June 21–22 and December 22–23.

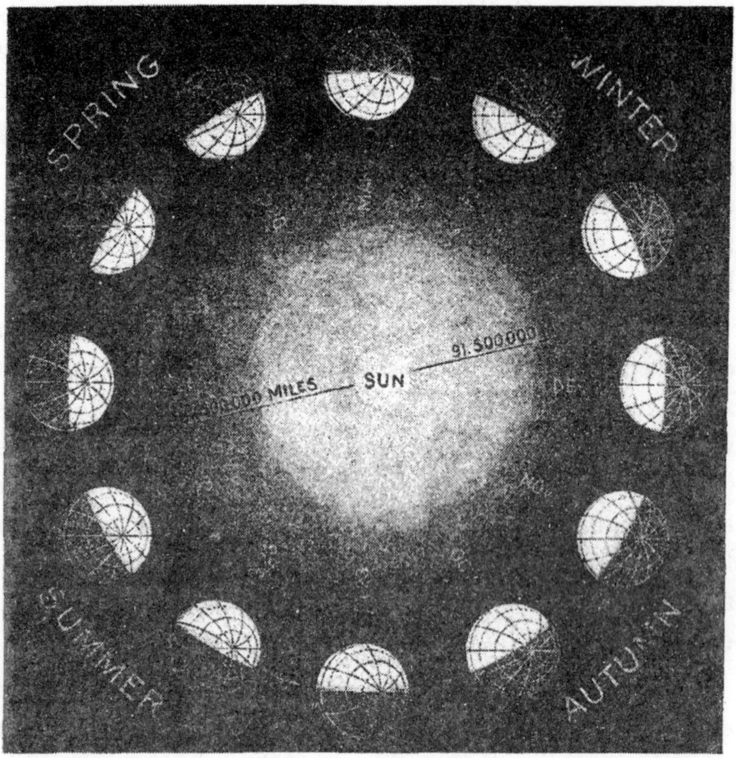

Fig. 45. Illumination of the earth in twelve positions, corresponding to months. The north pole is turned toward us.

it casts is due west and remains due west until noon, getting shorter and shorter as the sun rises higher.

Shadows. At noon the sun, being on the celestial equator, is directly overhead and casts no shadow, or the shadow is directly underneath. In the afternoon the shadow is due east, lengthening as the sun approaches the due west point in the horizon. At this time the sun's rays extend from pole to pole. *The circle of illumination*, that great circle separating the lighted half of the earth from the half which is turned away from the sun, since it extends at this time from pole to pole, coincides with a meridian circle and bisects each parallel. Half of each parallel being in the light and half in the dark, during one rotation every point will be in the light half a day and away from the sun the other half, and day and night are equal everywhere on the globe.

After March 21 the sun creeps back in its orbit, gradually, away from the celestial equator toward the North star. At the equator the sun thus rises more and more toward the north of the due east point on the horizon, and at noon casts a shadow toward the south. As the sun gets farther from the celestial equator, the south noon shadow lengthens, and the sun rises and sets farther toward the north of east and west.

On June 21 the sun has reached the point in the ecliptic farthest from the celestial equator, about 23½° north. The vertical ray on the earth is at a corresponding distance from the equator. The sun is near the constellation Cancer, and the parallel marking the turning of the sun from his course toward the polestar is called the Tropic (from a Greek word meaning turning) of Cancer. Our terrestrial parallel marking the southward turning of the vertical ray is also called the Tropic of Cancer. At this date the circle of illumination extends 23½° beyond the north pole, and all of the parallels north of 66½° from the equator are entirely within this circle of illumination and have daylight during the entire rotation of the earth. At this time the circle of illumination cuts unequally parallels north of the equator so that more than half of them are in the lighted portion, and hence days are longer than nights in the northern hemisphere. South of the equator the conditions are reversed. The circle of illumination does not extend so far south as the south pole, but falls short of it 23½°, and consequently all parallels south of 66½° are entirely in the dark portion of the earth, and it is continual night. Other circles south of the equator are so intersected by the circle of illumination that less than half of them are in the lighted side of the earth, and the days are shorter than the nights. It is midwinter there.

After June 21 gradually the sun creeps along in its orbit away from this northern point in the celestial sphere toward the celestial equator. The circle of illumination again draws toward the poles, the days are more nearly of the same length as the nights, the noon sun is more nearly overhead at the equator again, until by September 23, the autumnal equinox, the sun is again on the celestial equator, and conditions are exactly as they were at the March equinox.

After September 23 the sun, passing toward the South star from the celestial equator, rises to the south of a due east line on the equator, and at noon is to the south of the zenith, casting a north shadow. The circle of illumination withdraws from the north pole, leaving it in darkness, and extends beyond the south pole, spreading there the glad sunshine. Days grow shorter north of the equator, less than half of their parallels being in the lighted half, and south of the equator the days lengthen and summer comes.

On December 22 the sun has reached the most distant point in the ecliptic from the celestial equator toward the South star, 23½° from the celestial equator and 66½° from the South star, the vertical ray on the earth being at corresponding distances from the equator and the south pole. The sun is now near the constellation Capricorn, and everywhere within the tropics the shadow is toward the north; on the tropic of Capricorn the sun is overhead at noon, and south of it the shadow is toward the south. Here the vertical ray turns toward the equator again as the sun creeps in the ecliptic toward the celestial equator.

Just as the tropics are the parallels which mark the farthest limit of the vertical ray from the equator, the polar circles are the parallels marking the farthest extent of the circle of illumination beyond the poles, and are the same distance from the poles that the tropics are from the equator.

The Width of the Zones is thus determined by the distance the vertical ray travels on the earth, and with the moving of the vertical ray, the shifting of the day circle. This distance is in turn determined by the angle which the earth's orbit forms with the plane of the equator. The planes of the equator and the orbit forming an angle of 23½°, the vertical ray travels that many degrees each side of the equator, and the torrid zone is 47° wide. The circle of illumination never extends more than 23½° beyond each pole, and the frigid zones are thus 23½° wide. The remaining or temperate zones between the torrid and the frigid zones must each be 43° wide.

At the North Pole. Imagine you are at the north pole. Bear in mind the fact that the North star is always almost exactly overhead and the celestial equator always on the horizon. On March 21 the sun is on the celestial equator and hence on the horizon.* The sun now swings around the horizon once each rotation of the earth, casting long shadows in every direction, though, being at the north pole, they are always toward the south.† After the spring equinox, the sun

*Speaking exactly, the sun is seen there before the spring equinox and after the autumnal equinox, owing to refraction and the dip of the horizon. See p. 161.

† The student should bear in mind the fact that directions on the earth are determined solely by reference to the true geographical pole, not the magnetic pole of the mariner's compass. At the north pole the compass points due south, and at points between the magnetic pole and the geographical pole it may point in any direction excepting toward the north. Thus Admiral A. H. Markham says, in the *Youth's Companion* for June 22, 1902:

gradually rises higher and higher in a gently rising spiral until at the summer solstice, June 21, it is 23½° above the horizon. After this date it gradually approaches the horizon again until, September 23, the autumnal equinox, it is exactly on the horizon, and after this date is seen no more for six months. Now the stars come out and may be seen perpetually tracing their circular courses around the polestar. Because of the reflection and refraction of the rays of light in the air, twilight prevails when the sun is not more than about 18° below the horizon, so that for only a small portion of the six months' winter is it dark, and even then the long journeys of the moon above the celestial equator, the bright stars that never set, and the auroras, prevent total darkness (see p. 165). On December 22 the sun is 23½° below the horizon, after which it gradually approaches the horizon again, twilight soon setting in until March 21 again shows the welcome face of the sun.

At the South Pole the conditions are exactly reversed. There the sun swings around the horizon in the opposite direction; that is, in the direction opposite the hands of a watch when looked at from above. The other half of the celestial sphere from that seen at the north pole is always above one, and no stars seen at one pole are visible at the other pole, excepting the few in a very narrow belt around the celestial equator, lifted by refraction of light.

Parallelism of the Earth's Axis. Another condition of the earth in its revolution should be borne in mind in explaining change of seasons. The earth might rotate on an

"When, in 1876, I was sledging over the frozen sea in my endeavor to reach the north pole, and therefore traveling in a due north direction, I was actually steering by compass E. S. E., the variation of the compass in that locality varying from ninety-eight degrees to one hundred and two degrees westerly."

axis and revolve around the sun with the axis inclined 23½° and still give us no change in seasons. This can easily be demonstrated by carrying a globe around a central object representing the sun, and by rotating the axis one can maintain the same inclination but keep the vertical ray continually at the equator or at any other circle within the tropics. In order to get the shifting of the vertical ray and change of seasons which now obtain, the axis must constantly point in the same direction, and its position at one time be parallel to its position at any other time. This is called the parallelism of the earth's axis.

That the earth's axis has a very slow rotary motion, a slight periodic "nodding" which varies its inclination toward the plane of the ecliptic, and also irregular motions of diverse character, need not confuse us here, as they are either so minute as to require very delicate observations to determine them, or so slow as to require many years to show a change. These three motions of the axis are discussed in the Appendix under "Precession of the Equinoxes," "Nutation of the Poles," and "Wandering of the Poles" (p. 285).

Experiments with the Gyroscope. The *gyroscope*, probably familiar to most persons, admirably illustrates the causes of the parallelism of the earth's axis. A disk, supported in a ring, is rapidly whirled, and the rotation tends to keep the axis of the disk always pointing in the same direction. If the ring be held in the hands and carried about, the disk rapidly rotating, it will be discovered that any attempt to change the direction of the axis will meet with resistance. This is shown in the simple fact that a rapidly rotating top remains upright and is not easily tipped over; and, similarly, a bicycle running at a rapid rate remains erect, the rapid motion of the wheel (or top) giving the axis a tendency to remain in the same plane.

The gyroscope shown in Figure 46* is one used by Professor R. S. Holway of the University of California. It was made by mounting a six-inch sewing machine wheel on ball bearings in the fork of an old bicycle. Its advantages over those commonly used are its simplicity, the ball bearings, and its greater weight.

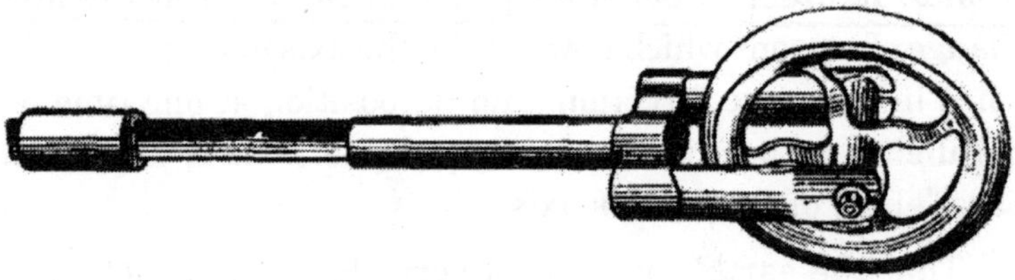

Fig. 46

Foucault Experiment. In 1852, the year after his famous pendulum experiment, demonstrating the rotation of the earth, M. Leon Foucault demonstrated the same facts by means of a gyroscope so mounted that, although the earth turned, the axis of the rotating wheel remained constantly in the same direction.

COMPARATIVE LENGTH OF DAY AND NIGHT

Day's Length at the Equinoxes. One half of the earth being always in the sunlight, the circle of illumination is a great circle. The vertical ray marks the center of the lighted half of the surface of the earth. At the equinoxes the vertical ray is at the equator, and the circle of illumination extends from pole to pole bisecting every parallel. Since at this time any given parallel is cut into two equal parts by the circle of illumination, one half of it is in the sunlight, and one half of it is in darkness, and during one rotation a point on a parallel

*Taken, by permission, from the Journal of Geography for February, 1904.

will have had twelve hours day and twelve hours night. (No allowance is made for refraction or twilight.)

Day's Length after the Equinoxes. After the vernal equinox the vertical ray moves northward, and the circle of illumination extends beyond the north pole but falls short of the south pole. Then all parallels, save the equator, are unequally divided by the circle of illumination, for more than half of each parallel north of the equator is in the light, and more than half of each parallel south of the equator is in darkness. Consequently, while the vertical ray is north of the equator, or from March 21 to September 23, the days are longer than the nights north of the equator, but are shorter than the nights south of the equator.

During the other half of the year, when the vertical ray is south of the equator, these conditions are exactly reversed. The farther the vertical ray is from the equator, the farther is the circle of illumination extended beyond one pole and away from the other pole, and the more unevenly are the parallels divided by it; hence the days are proportionally longer in the hemisphere where the vertical ray is, and the nights longer in the opposite hemisphere. The farther from the equator, too, the greater is the difference, as may be observed from Figure 50, page 163. Parallels near the equator are always nearly bisected by the circle of illumination, and hence day nearly equals night there the year around.

Day's Length at the Equator. How does the length of day at the equator compare with the length of night? When days are shorter south of the equator, if they are longer north of it and *vice versa*, at the equator they must be of the same length. The equator is always bisected by the circle of illumination, consequently half of it is always in the sunlight. This proposition, simple though it is, often needs further demonstration to be seen clearly. It will be obvious if one sees:

(*a*) A point on a sphere 180° in any direction from a point in a great circle lies in the same circle.

(*b*) Two great circles on the same sphere must cross each other at least once.

(*c*) A point 180° from this point of intersection common to both great circles, will lie in each of them, and hence must be a point common to both and a point of intersection. Hence two great circles, extending in any direction, intersect each other a second time 180° from the first point of crossing, or half way around. The circle of illumination and equator are both great circles and hence bisect each other. If the equator is always bisected by the circle of illumination, half of it must always be in the light and half in the dark.

Day's Length at the Poles. The length of day at the north pole is a little more than six months, since it extends from March 21 until September 23, or 186 days. At the north pole night extends from September 23 until March 21, and is thus 179 days in length. It is just opposite at the south pole, 179 days of sunshine and 186 days of twilight and darkness. This is only roughly stated in full days, and makes no allowance for refraction of light or twilight.

Longest Days at Different Latitudes. The length of the longest day, that is, from sunrise to sunset, in different latitudes is as follows:

Lat.	Day	Lat.	Day	Lat.	Day	Lat.	Day
0°	12 h.	25°	13 h. 34 m.	50°	16 h. 9 m.	70°	65 days
5°	12 h. 17 m.	30°	13 h. 56 m.	55°	17 h. 7 m.	75°	103 "
10°	12 h. 35 m.	35°	14 h. 22 m.	60°	18 h. 30 m.	80°	134 "
15°	12 h. 53 m.	40°	14 h. 51 m.	65°	21 h. 09 m.	85°	161 "
20°	13 h. 13 m.	45°	15 h. 26 m.	66° 33'	24 h. 00 m.	90°	6 mos.

The foregoing table makes no allowance for the fact that the vertical ray is north of the equator for a longer time than it is south of the equator, owing to the fact that we are farther from the sun then, and consequently the earth revolves more slowly in its orbit. No allowance is made for refraction, which lifts up the rays of the sun when it is near the horizon, thus lengthening days everywhere.

REFRACTION OF LIGHT

Refraction of Light

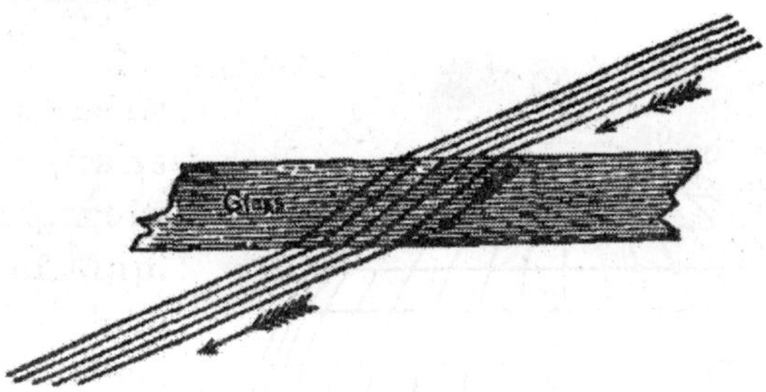

Fig. 47

The rays of light on entering the atmosphere are bent out of straight courses. Whenever a ray of light enters obliquely a medium of greater or of less density, the ray is bent out of its course (Fig. 47). Such a change in direction is called refraction. When a ray of light enters obliquely a medium of greater density, as in passing through from the upper rarer atmosphere to the lower denser layers, or from air into water, the rays are bent in the direction toward a perpendicular to the surface or less obliquely. This is called the first law of refraction. The second law of refraction is the converse of this; that is, on entering a rarer medium the ray is bent more obliquely or away from a perpendicular to the surface. When a ray of light from an object strikes the eye, we see the object in the direction taken by the ray as it enters the eye, and if the ray is refracted this will not be the real position of the object. Thus a fish in the water (Fig. 48) would see the adjacent boy as though the boy were nearly above it, for the ray from the boy to the fish is bent downwards, and the ray as it enters the eye of the fish seems to be coming from a place higher up.

Fig. 48

Amount of Refraction Varies. The amount of refraction depends upon the difference in the density of the media and the obliqueness with which the rays enter. Rays entering perpendicularly are not refracted at all.

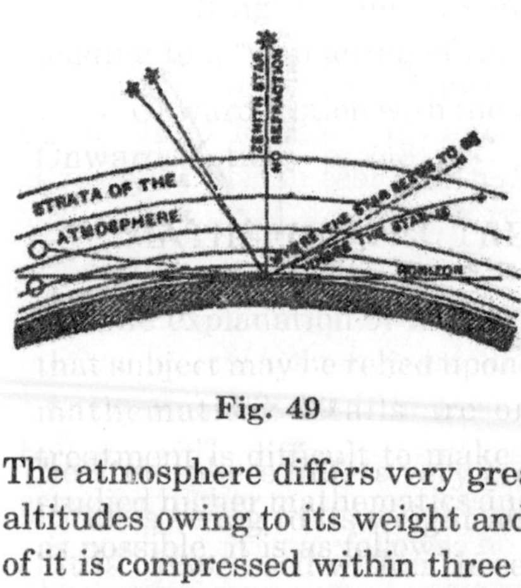

Fig. 49

The atmosphere differs very greatly in density at different altitudes owing to its weight and elasticity. About one half of it is compressed within three miles of the surface of the earth, and at a height of ten miles it is so rare that sound can scarcely be transmitted through it. A ray of light entering the atmosphere obliquely is thus obliged to traverse layers of air of increasing density, and is refracted more and more as it approaches the earth.

MEAN REFRACTION TABLE

(For Temperature 50° Fahr., barometric pressure 30 in.)

Apparent Altitude.	Mean Refraction.		Apparent Altitude.	Mean Refraction.		Apparent Altitude.	Mean Refraction.	
0°	36'	29.4"	8°	6'	33.3"	26°	1'	58.9"
1	24	53.6	9	5	52.6	30	1	40.6
2	18	25.5	10	5	19.2	40	1	9.4
3	14	25.1	12	4	27.5	50	0	48.9
4	11	44.4	14	3	49.5	60	0	33.6
5	9	52.0	16	3	20.5	70	0	21.2
6	8	28.0	18	2	57.5	80	0	10.3
7	7	23.8	22	2	23.3	90	0	00.0

Effect of Refraction on Celestial Altitudes. Thus, refraction increases the apparent altitudes of all celestial objects excepting those at the zenith (Fig. 49). The amount of refraction at the horizon is ordinarily 36′ 29″; that is to say, a star seen on the horizon is in reality over one half a degree below the horizon. The actual amount of refraction varies with the temperature, humidity, and pressure of the air, all of which affect its density and which must be taken into consideration in accurate calculations. Since the width of the sun as seen from the earth is about 32′, when the sun is seen just above the horizon it actually is just below it, and since the sun passes one degree in about four minutes, the day is thus lengthened about four minutes in the latitudes of the United States and more in higher latitudes. This accounts for the statement in almanacs as to the exact length of the day at the equinoxes. Theoretically the day is twelve hours long then, but practically it is a few minutes longer. Occasionally there is an eclipse of the moon observed just before the sun has gone down. The earth is exactly between the sun and the moon, but because of refraction, both sun and moon are seen above the horizon.

The sun and moon often appear attened when near the horizon, especially when seen through a haze. This apparent attening is due to the fact that rays from the lower portion are more oblique than those from the upper portion, and hence it is apparently lifted up more than the upper portion.

TWILIGHT

The atmosphere has the peculiar property of reflecting and scattering the rays of light in every direction. Were not this the case, no object would be visible out of the direct sunshine, shadows would be perfectly black, our houses, excepting where the sun shone, would be perfectly dark, the blue sky would disappear and we could see the stars in the day time just as well as at night. Because of this diffusion of light, darkness does not immediately set in after sunset, for the rays shining in the upper air are broken up and reflected to the lower air. This, in brief, is the explanation of twilight. There being practically no atmosphere on the moon there is no twilight there. These and other consequences resulting from the lack of an atmospheric envelope on the moon are described on p. 263.

Length of Twilight. Twilight is considered to last while the sun is less than about 18° below the horizon, though the exact distance varies somewhat with the condition of the atmosphere, the latitude, and the season of the year. There is thus a twilight zone immediately beyond the circle of illumination, and outside of this zone is the true night. Figure

50 represents these three portions: (1) the hemisphere receiving direct rays (slightly more than a hemisphere owing to refraction), (2) the belt 18° from the circle of illumination, and (3) the segment in darkness—total save for starlight or moonlight. The height of the atmosphere is, of course, greatly exaggerated. The atmosphere above the line *AB* receives direct rays of light and reflects and diffuses them to the lower layers of atmosphere.

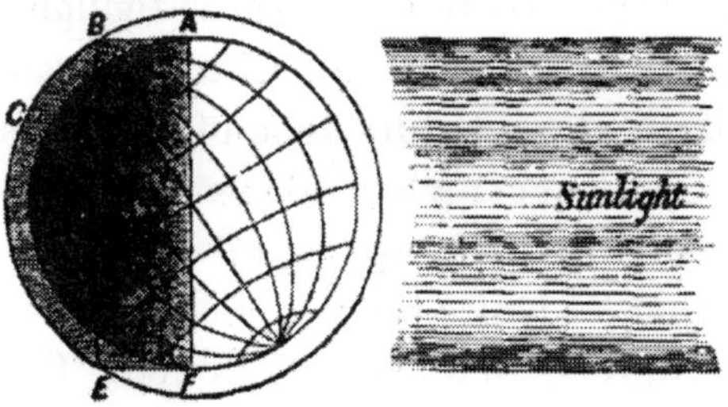

Fig. 50

Twilight Period Varies with Season. It will be seen from Figure 50 that the fraction of a parallel in the twilight zone varies greatly with the latitude and the season. At the equator the sun drops down at right angles to the horizon, hence covers the 18° twilight zone in $\frac{18}{360}$ of a day or one hour and twelve minutes. This remains practically the same the year around there. In latitudes of the United States, the twilight averages one and one-half hours long, being greater in midsummer. At the poles, twilight lasts about two and one-half months.

Twilight Long in High Latitudes. The reason why the twilight lasts so long in high latitudes in the summer will be apparent if we remember that the sun, rising north of east, swinging slantingly around and setting to the north of west, passes through the twilight zone at the same oblique angle. At latitude 48° 33′ the sun passes around so obliquely at the summer solstice that it does not sink 18° below the horizon at midnight, and stays within the twilight zone from sunset to sunrise. At higher latitudes on that date the sun sinks even less distance below the horizon. For example, at

St. Petersburg, latitude 59° 56′ 30″, the sun is only 6° 36′ 25″ below the horizon at midnight June 21 and it is light enough to read without artificial light. From 66° to the pole the sun stays entirely above the horizon throughout the entire summer solstice, that being the boundary of the "land of the midnight sun."

Twilight Near the Equator. "Here comes science now taking from us another of our cherished beliefs—the wide superstition that in the tropics there is almost no twilight, and that the 'sun goes down like thunder out o' China 'crosst the bay.' Every boy's book of adventure tells of travelers overtaken by the sudden descent of night, and men of science used to bear out these tales. Young, in his 'General Astronomy,' points out that 'at Quito the twilight is said to be at best only twenty minutes.' In a monograph upon 'The Duration of Twilight in the Tropics,' S. I. Bailey points out, by carefully verified observation and experiments, that the tropics have their fair share of twilights. He says: 'Twilight may be said to last until the last bit of illuminated sky disappears from the western horizon. In general it has been found that this occurs when the sun has sunk about eighteen degrees below the horizon... Arequipa, Peru, lies within the tropics, and has an elevation of 8,000 feet, and the air is especially pure and dry, and conditions appear to be exceptionally favorable for an extremely short twilight. On Sunday, June 25, 1899, the following observations were made at the Harvard Astronomical Station, which is situated here: The sun disappeared at 5.30 p.m., local mean time. At 6 p.m., thirty minutes after sunset, I could read ordinary print with perfect ease. At 6.30 p.m. I could see the time readily by an ordinary watch. At 6.40 p.m., seventy minutes after sunset, the illuminated western sky was still bright enough to cast a faint shadow of an opaque body on a white surface.

At 6.50 p.m., one hour and twenty minutes after sunset, it had disappeared. On August 27, 1899, the following observations were made at Vincocaya. The latitude of this place is about sixteen degrees south and the altitude 14, 360 feet. Here it was possible to read coarse print forty-seven minutes after sunset, and twilight could be seen for an hour and twelve minutes after the sun's disappearance.' So the common superstition about no twilight in the tropics goes to join the William Tell myth."—*Harper's Weekly*, April 5, 1902.

Twilight Near the Pole. "It may be interesting to relate the exact amount of light and darkness experienced during a winter passed by me in the Arctic regions within four hundred and sixty miles of the Pole.

"From the time of crossing the Arctic circle until we established ourselves in winter quarters on the 3d of September, we rejoiced in one long, continuous day. On that date the sun set below the northern horizon at midnight, and the daylight hours gradually decreased until the sun disappeared at noon below the southern horizon on the 13th of October.

"From this date until the 1st of March, a period of one hundred and forty days, we never saw the sun; but it must not be supposed that because the sun was absent we were living in total darkness, for such was not the case. During the month following the disappearance of the sun, and for a month prior to its return, we enjoyed for an hour, more or less, on either side of noon, a glorious twilight; but for three months it may be said we lived in total darkness, although of course on fine days the stars shone out bright and clear, rendered all the more brilliant by the reflection from the snow and ice by which we were surrounded, while we also enjoyed the light from the moon in its regular lunations.

"On the 21st of December, the shortest day in the year, the sun at our winter quarters was at noon twenty degrees below the horizon. I mention this because the twilight circle, or, to use its scientific name, the *crepusculum*, when dawn begins and twilight ends, is determined when the sun is eighteen degrees below the horizon.

"On our darkest day it was not possible at noon to read even the largest-sized type."—Admiral A. H. Markham, R. N., in the *Youth's Companion*, June 22, 1899.

EFFECT OF THE SHIFTING RAYS OF THE SUN.

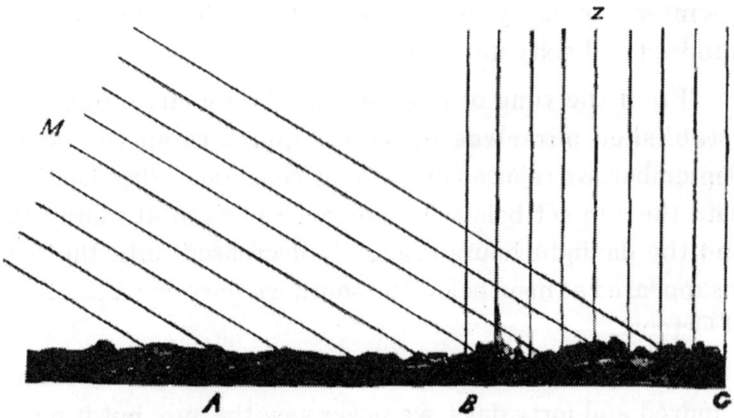

Fig. 51

Vertical Rays and Insolation. The more nearly vertical the rays of the sun are the greater is the amount of heat imparted to the earth at a given place, not because a vertical ray is any warmer, but because more rays fall over a given area. In Figure 51 we notice that more perpendicular rays extend over a given area than slanting ones. We observe the morning and evening rays of the sun, even when falling perpendicularly upon an object, say through a convex lens or burning glass, are not so warm as those at midday. The reason is apparent from Figure 52, the slanting rays traverse through more of the atmosphere.

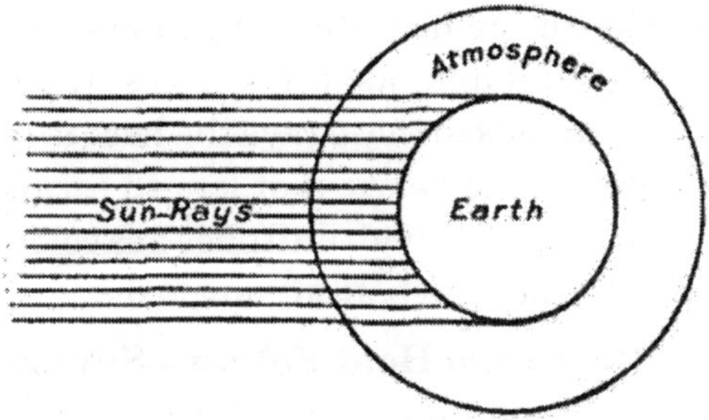

Fig. 52

At the summer solstice the sun's rays are more nearly vertical over Europe and the United States than at other times. In addition to the greater amount of heat received because of the less oblique rays, the days are longer than nights and consequently more heat is received during the day than is radiated off at night. This increasing length of day time greatly modifies the climate of regions far to the north. Here the long summer days accumulate enough heat to mature grain crops and forage plants. It is interesting to note that in many northern cities of the United States the maximum temperatures are as great as in some southern cities.

How the Atmosphere is Heated. To understand how the atmosphere gets its heat we may use as an illustration the peculiar heat-receiving and heat-transmitting properties of glass. We all know that glass permits heat rays from the sun to pass readily through it, and that the dark rays of heat from the stove or radiator do not readily pass through the glass. Were it not for this fact it would be no warmer in a room in the sunshine than in the shade, and if glass permitted heat to escape from a room as readily as it lets the sunshine in we should have to dispense with windows in cold weather. Stating this in more technical language, transparent glass is diathermanous to luminous heat rays but athermanous to dark rays. Dry air possesses this same peculiar property and permits the luminous rays from the sun to pass readily through to the earth, only about one fourth being absorbed as they pass through. About three fourths of the heat the atmosphere receives is that which is radiated back as dark rays from the earth. Being athermanous to these rays the heat is retained a considerable length of time before it at length escapes into space. It is for this reason that high altitudes are cold, the atmosphere being heated from the bottom upwards.

Maximum Heat Follows Summer Solstice. Because of these conditions and of the convecting currents of air, and, to a very limited extent, of water, the heat is so distributed and accumulated that the hottest weather is in the month following the summer solstice (July in the northern hemisphere, and January in the southern); conversely, the coldest month is the one following the winter solstice. This seasonal variation is precisely parallel to the diurnal change. At noon the sun is highest in the sky and pours in heat most rapidly, but the point of maximum heat is not usually reached until the middle of the afternoon, when the accumulated heat in the atmosphere begins gradually to disappear.

Astronomical and Climatic Seasons. Astronomically there are four seasons each year: spring, from the vernal equinox to the summer solstice; summer, from the summer solstice to the autumnal equinox; autumn, from the autumnal equinox to the winter solstice; winter, from the winter solstice to the spring equinox. As treated in physical geography, seasons vary greatly in number and length with differing conditions of topography and position in relation to winds, mountains, and bodies of water. In most parts of continental United States and Europe there are four fairly marked seasons: March, April, and May are called spring months; June, July, and August, summer months;

September, October, and November, autumn months; and December, January, and February, winter months. In the southern states and in western Europe the seasons just named begin earlier. In California and in most tropical regions, there are two seasons, one wet and one dry. In northern South America there are four seasons,—two wet and two dry.

From the point of view of mathematical geography there are four seasons having the following lengths in the northern hemisphere:

Spring:	Vernal equinox	March 21	92 days	Summer half 186 days
	Summer solstice.....	June 21		
Summer:	Summer solstice.....	June 21	94 days	
	Autumnal equinox	Sept. 23		
Autumn:	Autumnal equinox	Sept. 23	90 days	Winter half 179 days
	Winter solstice.....	Dec. 22		
Winter:	Winter solstice.....	Dec. 22	89 days	
	Vernal equinox	March 21		

Hemispheres Unequally Heated. For the southern hemisphere, spring should be substituted for autumn, and summer for winter. From the foregoing it will be seen that the northern hemisphere has longer summers and shorter winters than the southern hemisphere. Since the earth is in perihelion, nearest the sun, December 31, the earth as a whole then receives more heat than in the northern summer when the earth is farther from the sun. Though the earth as a whole must receive more heat in December than in July, the northern hemisphere is then turned away from the sun and has its winter, which is thus warmer than it would otherwise be. The converse is true of the northern summer. The earth then being in aphelion receives less heat each day, but the northern hemisphere being turned toward the sun then has its summer, cooler than it would be were this to occur when the earth is in perihelion. It is well to remember, however, that while the earth as a whole receives more heat in the half year of perihelion, there are only 179 days in that portion, and in the cooler portion there are 186 days, so that the total amount of heat received in each portion is exactly the same. (See Kepler's Second Law, p. 283.)

DETERMINATION OF LATITUDE FROM SUN'S MERIDIAN ALTITUDE.

In Chapter II we learned how latitude is determined by ascertaining the altitude of the celestial pole. We are now in a position to see how this is commonly determined by reference to the noon sun.

Relative Positions of Celestial Equator and Celestial Pole. The meridian altitude of the celestial equator at a given place and the altitude of the celestial pole at that place are complementary angles, that is, together they equal 90°. Though when understood this proposition is exceedingly simple, students sometimes only partially comprehend it, and the later conclusions are consequently hazy.

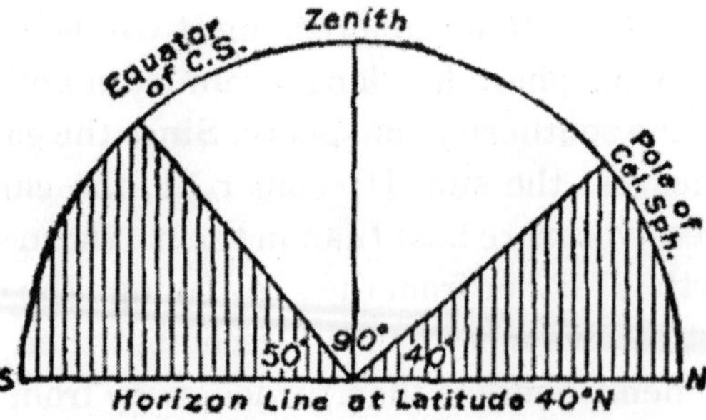

Fig. 53

1. The celestial equator is always 90. from the celestial pole.

2. An arc of the celestial sphere from the northern horizon through the zenith to the southern horizon comprises 180°.

3. Since there are 90. from the pole to the equator, from the northern horizon to the pole and from the southern horizon to the equator must together equal 90°.

One of the following statements is incorrect. Find which one it is.

a. In latitude 30° the altitude of the celestial pole is 30° and that of the celestial equator is 60°.

b. In latitude 36° the altitude of the celestial equator is 54°.

c. In latitude 48° 20′ the altitude of the celestial equator is 41° 40′.

d. If the celestial equator is 51° above the southern horizon, the celestial pole is 39. above the northern horizon.

e. If the altitude of the celestial equator is 49° 31′, the latitude must be 40° 29°.

f. If the altitude of the celestial equator is 21° 24′, the latitude is 69° 36′.

On March 21 the sun is on the celestial equator.* If on this day the sun's noon shadow indicates an altitude of 40° we know that is the altitude of the celestial equator, and this subtracted from 90° equals 50° the latitude of the place. On September 23 the sun is again on the celestial equator, and its noon altitude subtracted from 90° equals the latitude of the place where the observation is made.

*Of course, the center of the sun is not on the celestial equator all day, it is there but the moment of its crossing. The vernal equinox is the point of crossing, but we commonly apply the term to the day when the passage of the sun's center across the celestial equator occurs. During this day the sun travels northward less than 24′, and since its diameter is somewhat more than 33′ some portion of the sun's disk is on the celestial equator the entire day.

Declination of the Sun. The declination of the sun or of any other heavenly body is its distance north or south of the celestial equator. The analemma, shown on page 128, gives the approximate declination of the sun for every day in the year. The Nautical Almanac, Table 1, for any month gives the declination very exactly (to the tenth of a second) at apparent sun noon at the meridian of Greenwich, and the difference in declination for every hour, so the student can get the declination at his own longitude for any given day very exactly from this table. Without good instruments, however, the proportion of error of observation is so great that the analemma will answer ordinary purposes.

Fig. 54

How to Determine the Latitude of Any Place. By ascertaining the noon altitude of the sun, and referring to the analemma or a declination table, one can easily compute the latitude of a place.

1. First determine when the sun will be on your meridian and its shadow strike a north-south line. This is discussed on p. 129.

2. By some device measure the altitude of the sun at apparent noon; i.e., when the shadow is north. A cardboard placed level under a window shade, as illustrated in Figure 54, will give surprisingly accurate results; a carefully mounted quadrant (see Fig. 55),

however, will give more uniformly successful measurements. Angle *A* (Fig. 54), the shadow on the quadrant, is the altitude of the sun. This is apparent from Figure 56, since *xy* is the line to the sun, and angle *B* = angle *A*.

3. Consult the analemma and ascertain the declination of the sun. Add this to the sun's altitude if south declination, and subtract it if north declination. If you are south of the equator you must subtract declination south and add declination north. (If the addition makes the altitude of the sun more than 90°, subtract 90° from it, as under such circumstances you are north of the equator if it is a south shadow, or south of the equator if it is a north shadow. This will occur only within the tropics.)

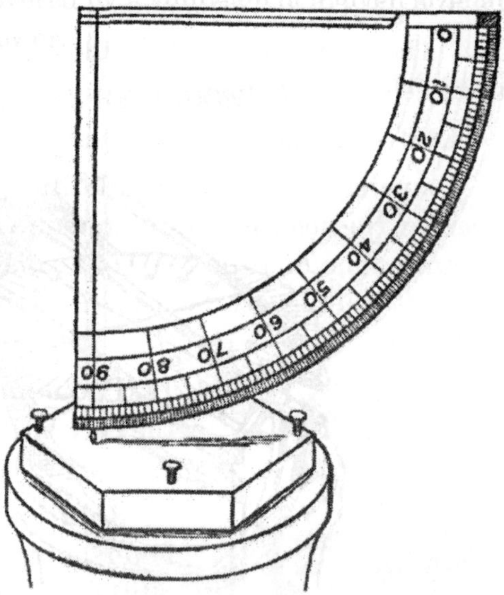

Fig. 55

4. Subtract the result of step three from 90°, and the remainder is your latitude.

Example. For example, say you are at San Francisco, October 23, and wish to ascertain your latitude.

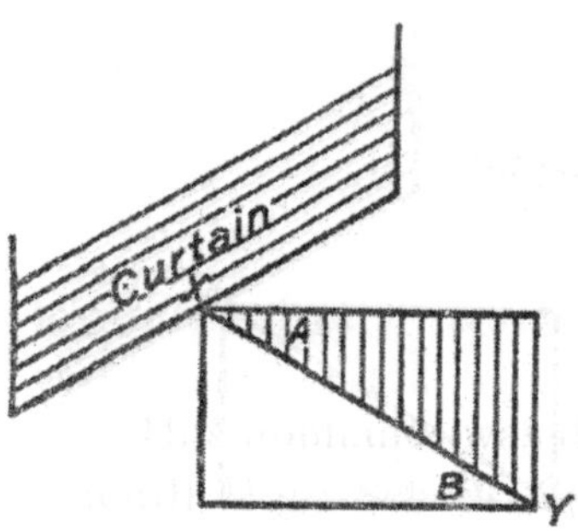

Fig. 56

1. Assume you have a north-south line. (The sun's shadow will cross it on that date at 11 h. 54 m. 33 s., a.m., Pacific time.)

2. The altitude of the sun when the shadow is north is found to be 41°.

3. The declination is 11° S. Adding we get 52°, the altitude of the celestial equator.

4. 90° - 52° equals 38°, latitude of place of observer.

Conversely, knowing the latitude of a place, one can ascertain the noon altitude of the sun at any given day. From the analemma and the table of latitudes many interesting problems will suggest themselves, as the following examples illustrate.

Fig. 57. Taking the altitude of the sun at sea

Problem. 1. How high above the horizon does the sun get at St. Petersburg on December 22?

Solution. The latitude of St. Petersburg is 59° 56′ N., hence the altitude of the celestial equator is 30.°40′. The declination of the sun December 22 is 23°27′ S. Since south is below the celestial equator at St. Petersburg, the altitude of the sun is 30° 4′ less 23 ° 27′, or 6 ° 37′

Problem. 2. At which city is the noon sun higher on June 21, Chicago or Quito?

Solution. The latitude of Chicago is 41 ° 5′ and the altitude of the celestial equator, 48° 1′. The declination of altitude of the sun at the sun June 21 is 23° 27′ N. North be-sea ing higher than the celestial equator at Chicago, the noon altitude of the sun is 48° 1′. plus 23° 27′, or 71° 37′.

The latitude of Quito being 0 °, the altitude of the celestial equator is 90°. The declination of the sun being 23° 27′ from this, the sun's noon altitude must be 90° less 23° 27′ or 66 ° 33′. The sun is thus 5° 4′ higher at Chicago than at the equator on June 21.

Latitude from Moon or Stars. With a more extended knowledge of astronomy and mathematics and with suitable instruments, we might ascertain the position of the celestial equator in the morning or evening from the moon, planets, or stars as well as from the sun. At sea the latitude is commonly ascertained by making measurements of the altitudes of the sun at apparent noon with the sextant. The declination tables are used, and allowances are made for refraction and for the "dip" of the horizon, and the resultant calculation usually gives the latitude within about half a mile. At observatories, where the latitude must be ascertained with the minutest precision possible, it is usually ascertained from star observations with a zenith telescope or a "meridian circle" telescope, and is verified in many ways.

CHAPTER-IX

TIDES

Tides and the Moon. The regular rise and fall of the level of the sea and the accompanying inflows and outflows of streams, bays, and channels, are called tides. Since very ancient times this action of the water has been associated with the moon because of the regular interval elapsing between a tide and the passage of the moon over the meridian of the place, and a somewhat uniform increase in the height of the tide when the moon in its orbit around the earth is nearest the sun or is farthest from it. This unquestioned lunar inuence on the ocean has doubtless been responsible as the basis for thousands of unwarranted associations of cause and effect of weather, vegetable growth, and even human temperament and disease with phases of the moon or planetary or astral conditions.

Other Periodic Ebbs and Flows. Since there are other periodical ebbs and flows due to various causes, it may be well to remember that the term tide properly applies only to the periodic rise and fall of water due to unbalanced forces in the attraction of the sun and moon. Other conditions which give rise to more or less periodical ebbs and flows of the oceans, seas, and great lakes are:

a. Variation in atmospheric pressure; low barometer gives an uplift to water and high barometer a depression.

b. Variability in evaporation, rainfall and melting snows produces changes in level of adjacent estuaries.

c. Variability in wind direction, especially strong and continuous seasonal winds like monsoons, lowers the level on the leeward of coasts and piles it up on the windward side.

d. Earthquakes sometimes cause huge waves.

A few preliminary facts to bear in mind when considering the causes of tides:

THE MOON

Sidereal Month. The moon revolves around the earth in the same direction that the earth revolves about the sun, from west to east. If the moon is observed near a given star on one night, twenty-four hours later it will be found, on the average, about 13.2° to the eastward. To reach the same star a second time it will require as many days as that distance is contained times in 360° or about 27.3 days. This is the sidereal month, the time required for one complete revolution of the moon.

Synodic Month. Suppose the moon is near the sun at a given time, that is, in the same part of the celestial sphere. During the twenty-fours hours following, the moon will creep eastward 13.2° and the sun 1°. The moon thus gains on the suneach day about 12.2, and to get in conjunction with it a second time it will take as many days as 12.2° is contained in 360° or about 29.5 days. This is called a synodic (from a Greek word meaning "meeting") month, the time from conjunction with the sun—new moon—until the next conjunction or new moon. The term is also applied to the time from opposition or full moon until the next opposition or full moon. If the phases of the moon are not clearly understood it would be well to follow the suggestions on this subject in the first chapter.

Moon's Orbit. The moon's orbit is an ellipse, its nearest point to the earth is called perigee (from *peri*, around or near; and *ge*, the earth) and is about 221, 617 miles. Its most distant point is called apogee (from *apo*, from; and *ge*, earth) and is about 252, 972 miles. The average distance of the moon from the earth is 238, 840 miles. The moon's orbit is inclined to the ecliptic 5° 8' and thus may be that distance farther north or south than the sun ever gets.

The new moon is said to be in *conjunction* with the sun, both being on the same side of the earth. If both are then in the plane of the ecliptic an eclipse of the sun must take place. The moon being so small, relatively (diameter 2, 163 miles), its shadow on the earth is small and thus the eclipse is visible along a relatively narrow path.

The full moon is said to be in *opposition* to the sun, it being on the opposite side of the earth. If, when in opposition, the moon is in the plane of the ecliptic it will be eclipsed by the shadow of the earth. When the moon is in conjunction or in opposition it is said to be in *syzygy*.

GRAVITATION

Laws Restated. This force was discussed in the first chapter where the two laws of gravitation were explained and illustrated. The term gravity is applied to the force of gravitation exerted by the earth (see Appendix, p. 278). Since the explanation of tides is simply the application of the laws of gravitation to the earth, sun, and moon, we may repeat the two laws:

First law: The force of gravitation varies directly as the mass of the object.

Second law: The force of gravitation varies inversely as the square of the distance of the object.

Sun's Attraction Greater, but Moon's Tide-Producing Inuence Greater. There is a widely current notion that since the moon causes greater tides than the sun, in the ratio of 5 to 2, the moon must have greater attractive inuence for the earth than the sun has. Now this cannot be true, else the earth would swing around the moon as certainly as it does around the sun. Applying the laws of gravitation to the problem, we see that the sun's attraction for the earth is approximately 176 times that of the moon.*

The reasoning which often leads to the erroneous conclusion just referred to, is probably something like this:

Major premise: Lunar and solar attraction causes tides.

Minor premise: Lunar tides are higher than solar tides.

Conclusion: Lunar attraction is greater than solar attraction.

We have just seen that the conclusion is in error. One orboth of the premises must be in error also. A study of the causes of tides will set this matter right.

CAUSES OF TIDES†

It is sometimes erroneously stated that wind is caused by heat. It would be more nearly correct to say that wind is caused by the unequal heating of the atmosphere. Similarly, it is not the attraction of the sun and moon for the earth that causes tides, it is the unequal attraction for different portions of the earth that gives rise to unbalanced forces which produce tides.

*For the method of demonstration, see p. 19. The following data are necessary: Earth's mass, 1/81; sun's mass, 330, 000; moon's mass, distance of earth to sun, 93, 000, 000 miles; distance of earth to moon, 239, 000 miles.

†A mathematical treatment will be found in the Appendix.

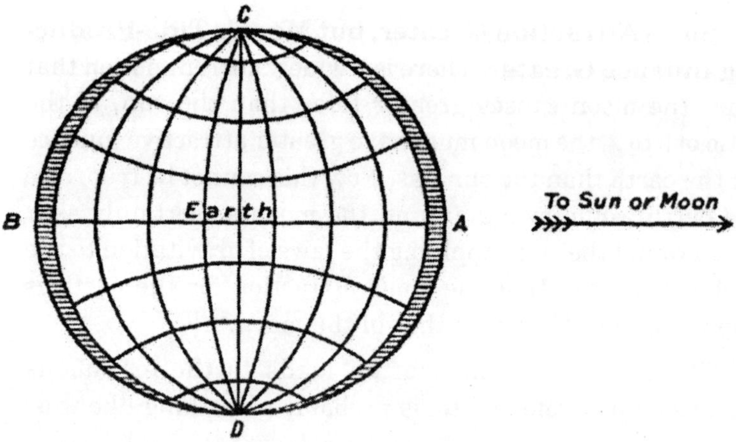

Fig. 58

Portions of the earth toward the moon or sun are 8, 000 miles nearer than portions on the side of the earth opposite the attracting body, hence the force of gravitation is slightly different at those points as compared with other points on the earth's surface. It is obvious, then, that at *A* and *B* (Fig. 58) there are two unbalanced forces, that is, forces not having counterparts elsewhere to balance them. At these two sides, then, tides are produced, since the water of the oceans yields to the inuence of these forces. That this may be made clear, let us examine these tides separately.

The Tide on the Side of the Earth Toward the Moon. If *A* is 239, 000 miles from the moon, *B* is 247, 000 miles away from it, the diameter of the earth being *AB* (Fig. 58). Now the attraction of the moon at *A, C,* and *D*, is away from the center of the earth and thus lessens the force of gravity at those points, lessening more at *A* since *A* is nearer and the moon's attraction is exerted in a line directly opposite to that of gravity. The water, being fluid and easily moved, yields to this lightening of its weight and tends to "pile up under the moon." We thus have a tide on the side of the earth toward the moon.

Tidal Wave Sweeps Westward. As the earth turns on its axis it brings successive portions of the earth toward the moon and this wave sweeps around the globe as nearly as possible under the moon. The tide is retarded somewhat by shallow water and the configuration of the coast and is not found at a given place when the moon is at meridian height but lags somewhat behind. The time between the passage of the moon and high tide is called the *establishment of the port.* This time varies greatly at different places and varies somewhat at different times of the year for the same place.

Solar Tides Compared with Lunar Tides. Solar tides are produced on the side of the earth toward the sun for exactly the same reason, but because the sun is so far away its attraction is more uniform upon different parts of the earth. If *A* is 93,000,000 miles from the sun, *B* is 93,008,000 miles from the sun. The ratio of the squares of these two numbers is much nearer unity than the ratio of the numbers representing the squares of the distances of *A* and *B* from the moon. If the sun were as near as the moon, the attraction for *A* would be greater by an enormous amount as compared with its attraction for *B*. Imagine a ball made of dough with lines connected to every particle. If we pull these lines uniformly the ball will not be pulled out of shape, however hard we pull. If, hwever, we pull some lines harder than others, although we pull gently, will not the ball be pulled out of shape? Now the pull of the sun, while greater than that of the moon, is exerted quite evenly throughout the earth and has but a slight tide-producing power. The attraction of the moon, while less than that of the sun, is exerted less evenly than that of the sun and hence produces greater tides.

It has been demonstrated that the tide-producing force of a body varies inversely as the cube of its distance and directly as its mass. Applying this to the moon and sun we get:

Let T = sun's tide-producing power,

and t = moon's tide-producing power.

The sun's mass is 26, 500, 000 times the moon's mass,

$$\therefore T : t :: 26,500,000 : 1.$$

But the sun's distance from the earth is 390 times the moon's distance,

$$\therefore T : t :: \frac{1}{390^3} : 1.$$

Combining the two proportions, we get,

$$T : t :: 2 : 5.$$

It has been shown that, owing to the very nearly equal attraction of the sun for different parts of the earth, a body's weight is decreased when the sun is overhead, as compared wih $\frac{1}{20,000,000}$ the weight six hours from then, by only ; that is, an object weighing a ton varies in weight ¾ of a grain from sunrise to noon. In case of the moon this difference is about 2½ times as great, or nearly 2 grains.

Tides on the Moon. It may be of interest to note that the effect of the earth's attraction on different sides of the moon must be twenty times as great as this, so it is thought that when the moon was warmer and had oceans* the tremendous tidal waves swinging around in the opposite

*The presence of oceans or an atmosphere is not essential to the theory, indeed, is not usually taken into account. It seems most certain that the earth is not perfectly rigid, and the theory assumes that the planets and the moon have sufficient viscosity to produce body tides.

direction to its rotation caused a gradual retardation of its rotation until, as ages passed, it came to keep the same face toward the earth. The planets nearest the sun, Mercury and Venus, probably keep the same side toward the sun for a similar reason. Applying the same reasoning to the earth, it is believed that the period of rotation must be gradually shortening, though the rate sems to be entirely inappreciable.

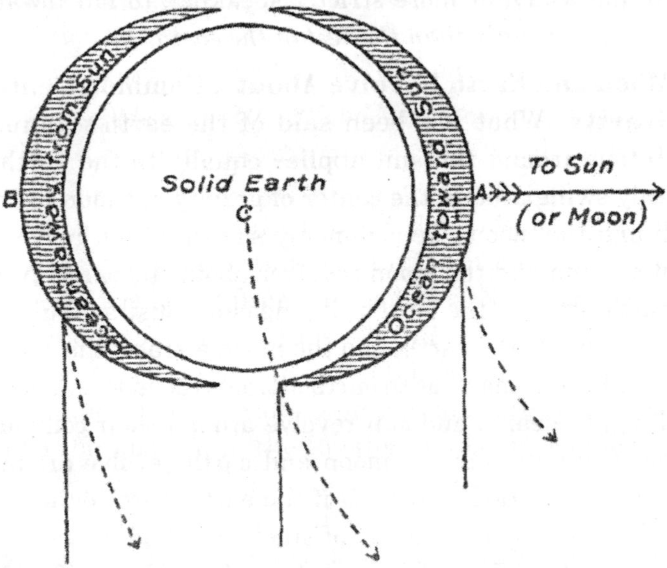

Fig. 59

The Tide on the Side of the Earth Opposite the Moon. A planet revolving around the sun, or a moon about a planet, takes a rate which varies in a definite mathematical ratio to its distance (see p. 284). The sun pulls the earth toward itself about one ninth of an inch every second. If the earth were nearer, its revolutionary motion would be faster. In case of planets having several satellites it is observed that the nearer ones revolve about the planet faster than the outer ones (see p. 255). Now if the earth were divided into three separate portions, as in Figure 59, the ocean nearest the sun, the earth proper, and the ocean opposite the sun would have three separate mtions somewhat as the dotted lines show. Ocean *A* would revolve faster than earth *C* or ocean *B*. If these three portions were connected by weak bands their stretching apart would cause them to separate entirely. The tide-producing power at *B* is this tendency it has to fall away, or more strictly speaking, to fall toward the sun *less rapidly than the rest of the earth.*

Moon and Earth Revolve About a Common Center of Gravity. What has been said of the earth's annual revolution around the sun applies equally to the earth's monthly swing around the center of gravity common to the earth and the moon. We commonly say the earth revolves about the sun and the moon revolves about the earth. Now the earth attracts the sun, in its measure, just as truly as the sun attracts the earth; and the moon attracts the earth, in the ratio of its mass, as th earth attracts the moon. Strictly speaking, the earth and sun revolve around their common center of gravity and the moon and earth revolve around their center of gravity. It is as if the earth were connected with the moon by a rigid bar of steel (that had no weight) and the two, thus firmly bound at the ends of this rod 239,000 miles long, were set spinning. If both were of the same weight, they would revolve about a point equidistant from each. The weight of the moon being somewhat less than $\frac{1}{81}$ that of the earth, this center of gravity, or point of balance, is only about 1,000 miles from the earth's center.

Spring Tides. When the sun and moon are in conjunction, both on the same side of the earth, the unequal attraction of both for the side toward them produces an unusually high tide there, and the increased centrifugal force at the side opposite them also produces an unusual high tide there. Both solar tides and both lunar tides are also combined when the sun and moon are in opposition. Since the sun and moon are in syzygy (opposition or conjunction) twice a month, high tides, called spring tides, occur at every new moon and at every full moon. If the moon should be in perigee, nearest the earth, at the same time it was new or full moon, spring tides would be unusually high.

Neap Tides. When the moon is at first or last quarter—moon, earth, and sun forming a right angle—the solar tides occur in the trough of the lunar tides and they are not as low as usual, and lunar tides occurring in the trough of the solar tides are not so high as usual.

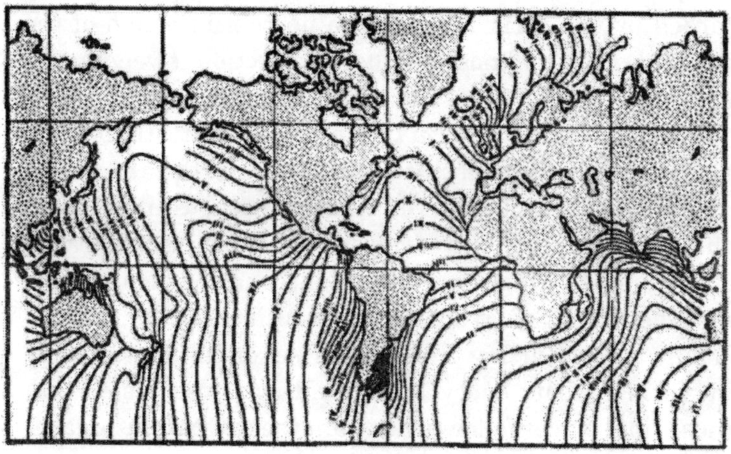

Fig. 60. Co-tidal lines

Course of the Tidal Wave. While the tidal wave is generated at any point under or opposite the sun or moon, it is out in the southern Pacific Ocean that the absence of shallow water and land areas offers least obstruction to its movement. Here a general lifting of the ocean occurs, and as the earth rotates the lifting progresses under or opposite the moon or sun from east to west. Thus a huge wave with crest extending north and south starts twice a day off the western coast of South America. The general position of this crest is shown on the co-tidal map, one line for every hour's difference in time. The tidal wave is retarded along its northern extremity, and as it sweeps along the coast of northern South America and North America, the wave assumes a northwesterly direction and sweeps down the coast of Asia at the rate of about 850 miles per hour. The southern portion passes across the Indian Ocean, being retarded in the north so that the southern portion is south of Africa when the northern portion has just reached southern India. The time it has taken the crest to pass from South America to south Africa is about 30 hours. Being retarded by the African coast, the northern portion of the wave assumes an almost northerly direction, sweeping up the Atlantic at the rate of about 700 miles an hour. It moves so much faster northward in the central Atlantic than along the coasts that the crest bends rapidly northward in the center and strikes all points of the coast of the United States within two or three hours of the same time. To reach France the wave must swing around Scotland and then southward across the North Sea, reaching the mouth of the Seine about 60 hours after starting from South America. A new wave being formed about every 12 hours, there are thus several of these tidal waves following one another across the oceans, each slightly different from the others.

While the term "wave" is correctly applied to this tidal movementit is very liable to leave a wrong impression upon the minds of those who have never seen the sea. When thinking of this tidal wave sweeping across the ocean at the rate of several hundred miles per hour, we should also bear in mind its height and length (by height is meant the vertical distance from the trough to the crest, and by length the distance from crest to crest). Out in midocean the height is only a foot or two and the length is hundreds of miles. Since the wave requires about three hours to pass from trough to crest, it is evident that a ship at sea is lifted up a foot or so during six hours and then as slowly lowered again, a motion not easily detected. On the shore the height is greater and the wave-length shorter, for about six hours the water gradually rises and then for about six hours it ebbs away again. Breakers, bores, and unusual tide phenomena are discussed on p. 188.

Time Between Successive Tides. The time elapsing from the passage of the moon across a meridian until it crosses the same meridian again is 24 hours 51 min.* This, in contradistinction to the solar and sidereal day, may be termed a lunar day. It takes the moon 27.3 solar days to revolve around the earth, a sidereal month. In one day it journeys $\frac{10}{273}$ of a day or 51 minutes. So if the moon was on a given meridian at 10 a.m., on one day, by 10 a.m. the next day the moon would have moved 12.2° eastward, and to direct the same meridian a second time toward the moon it takes on the average 51 minutes longer than 24 hours, the actual time varying from 38 m. to 1 h. 6 m. for various reasons. The tides of one day, then, are later than the tides of the preceding day by an average interval of 51 minutes.

In studying the movement of the tidal wave, we observed that it is retarded by shallow water. The spring tides being higher and more powerful move faster than the neap tides, the interval on successive days averaging only 38 minutes. Neap tides move slower, averaging somewhat over an hour

*More precisely, 24 h. 50 m. 51 s. This is the mean lunar day, or interval between successive passages of the moon over a given meridian. The apparent lunar day varies in length from 24 h. 38 m. to 25 h. 5 m. for causes somewhat similar to those producing a variation in the length of the apparent solar day.

Fig. 61. Low tide

later from day to day. The establishment of a port, as previously explained, is the average time elapsing between the passage of the moon and the high tide following it. The establishment for Boston is 11 hours, 27 minutes, although this varies half an hour at different times of the year.

Height of Tides. The height of the tide varies greatly in different places, being scarcely discernible out in midocean, averaging only 1½ feet in the somewhat sheltered Gulf of Mexico,but averaging 37 feet in the Bay of Fundy. The shape and situation of some bays and mouths of rivers is such that as the tidal wave enters, the front part of the wave becomes so steep that huge breakers form and roll up the bay or river with great speed. These bores, as they are called, occur in the Bay of Fundy, in the Hoogly estuary of the Ganges, in that of the Dordogne, the Severn, the Elbe, the Weser, the Yangtze, the Amazon, etc.

Bore of the Amazon. A description of the bore of the Amazon, given by La Condamine in the eighteenth century, gives a good idea of this phenomenon. "During three days before the new and full moons, the period of the highest tides, the sea, instead of occupying six hours to reach its ood, swells to its highest limit in one or two minutes. The noise of this terrible ood is heard five or six miles, and increases as it approaches. Presently you see a liquid promontory, 12 or 15 feet high, followed by another, and another, and sometimes by a fourth. These watery mountains spread across the whole channel, and advance with a prodigious rapidity, rending and crushing everything in their way. Immense trees are instantly uprooted by it, and sometimes whole tracts of land are swept away."

CHAPTER-X

MAP PROJECTIONS

To represent the curved surface of the earth, or any portion of it, on the plane surface of a map, involves serious mathematical difficulties. Indeed, it is impossible to do so with perfect accuracy. The term projection, as applied to the representation on a plane of points corresponding to points on a globe, is not always used in geography in its strictly mathematical sense, but denotes any representation on a plane of parallels and meridians of the earth.

Fig. 62

THE ORTHOGRAPHIC PROJECTION

This is, perhaps, the most readily understood projection, and is one of the oldest known, having been used by the ancient Greeks for celestial representation. The globe truly represents the relative positions of points on the earth's surface. It might seem that a photograph of a globe would correctly represent on a at surface the curved surface of the earth. A glance at Figure 62, from a photograph of a globe, shows the parallels near the equator to be farther apart than those near the poles. This is not the way they are on the globe. The orthographic projection is the representation of the globe as a photograph would show it from a great distance.

Parallels and Meridians FartherApart in Center of Map. Viewing a globe from a distance, we observe that parallels and meridians appear somewhat as represented in Figure 63,being farther apart toward the center and increasingly nearer toward the outer portion. Now it is obvious from Figure 64 that the farther the eye is projection placed from the globe, the less will be the distortion, although a removal to an infinite distance will not obviate all distortion.

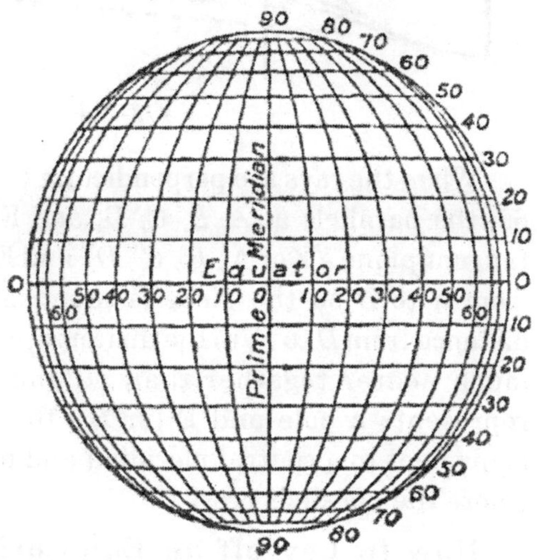

Fig. 63. Equatorial orthographic projection

Thus the eye at x sees lines to E and F much nearer together than lines to A and B, but the eye at the greater distance sees less difference.

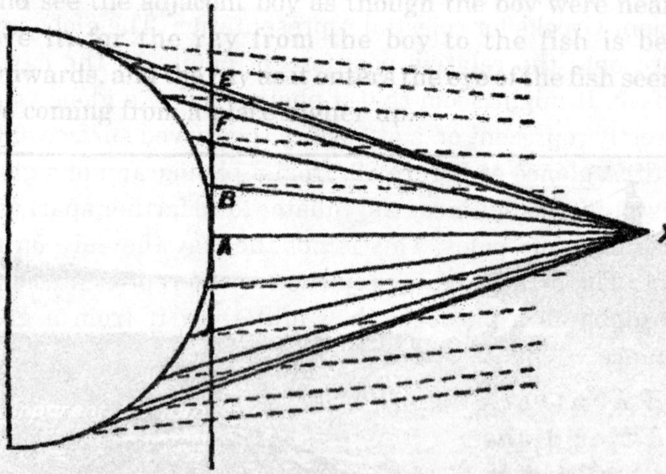

Fig. 64

When the rays are perpendicular to the axis, as in Figure 65, the parallels at *A, B, C, D*, and E will be seen on the tangent plane *XY* at *A'*, *B'*, *C'*, *D'*, and E'. While the distance from *A* to *B* on the globe is practically the same as the distance from *D* to *E*, to the distant eye *A'* and *B'* will appear much nearer together than *D'* and *E'*. Since *A* (or *A'*) represents a pole and E (or *E'*) the equator, line *XY* is equivalent to a central meridian and points *A'*, *B'*, etc., are where the parallels cross it.

How to Lay off an Equatorial Orthographic Projection. If parallels and meridians are desired for every 15°, divide the circle into twenty-four equal parts; any desired number of parallels and meridians, of course, may be drawn. Now connect opposite points with straight lines for parallels (as in Fig. 65). The reason why parallels are straight lines in the equatorial orthographic projection is apparent if one remembers that if the eye is in the plane of the equator and is at an infinite distance, the parallels will lie in practically the same plane as the eye.

To lay off the meridians, mark on the equator points exactly as on the central meridian where parallels intersect it. The meridians may now be made as arcs of circles passing through the poles and these points. With one foot of the compasses in the equator, or equator extended, place the other so that it will pass through the poles and one of these points. After a little trial it will be easy to lay off each of the meridians in this way.

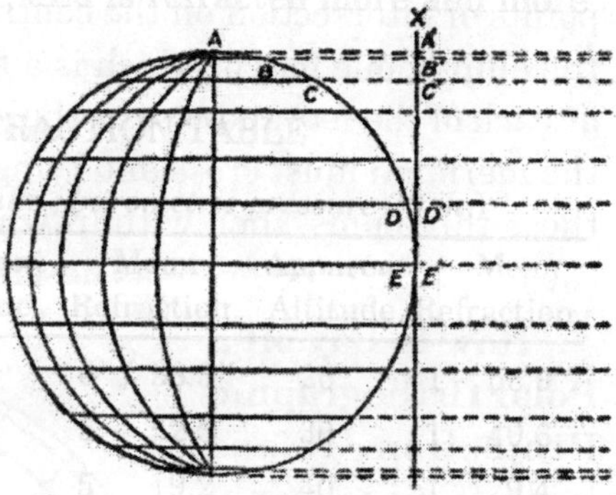

Fig. 65

To be strictly correct the meridians should not be arcs of circles as just suggested but should be semi-ellipses with the central meridian as major axis as shown in Figure 66. While somewhat more difficult, the student should learn how thus to lay them off. To construct the ellipse, one must first locate the foci. This is done by taking half the major axis (central meridian)

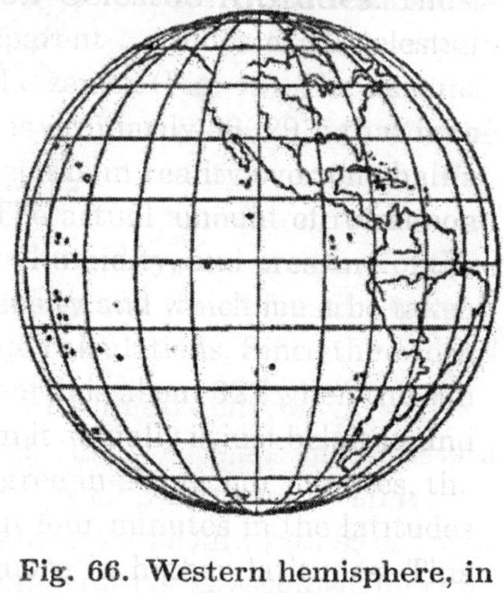

Fig. 66. Western hemisphere, in equatorial orthographic projection

as radius and with the point on the equator through which the meridian is to be constructed as center, describe an arc cutting the center meridian on each side of the equator. These points of intersection on the central meridian are the foci of the ellipse, one half of which is a meridian. By placing a pin at each of the foci and also at the point in the equator where the meridian must cross and tying a string as a loop around these three pins, then withdrawing the one at the equator, the ellipse may be made as described in the first chapter.

How to Lay off a Polar Orthographic Projection. This is laid off more easily than the former projection. Here the eye is conceived to be directly above a pole and the equator is the boundary of the hemisphere seen. It is apparent that from this position the equator and parallels will appear as circles and, since the planes of the meridians pass through the eye, each meridian must appear as a straight line.

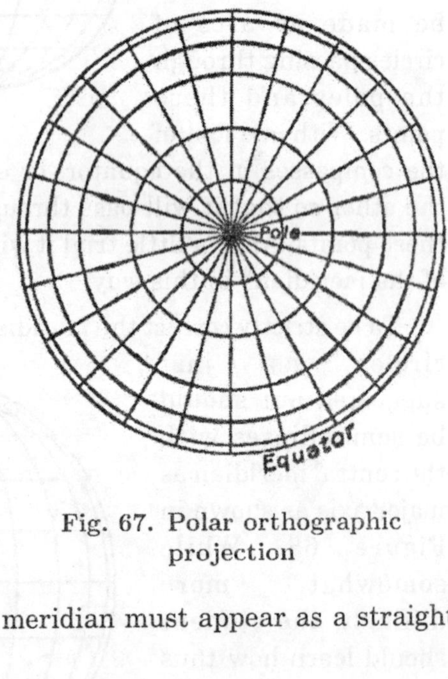

Fig. 67. Polar orthographic projection

Lay off for the equator a circle the same size as the preceding one (Fig. 65), subdividing it into twenty-four parts, if meridians are desired for every 15° Connect these points with the center, which represents the pole. On any diameter mark off distances as on the center meridian of the equatorial orthographic projection (Fig. 65). Through these points draw circles to repre sent parallels. You will then have the complete projection as in Figure 67.

Projections may be made with any point on the globe as center, though the limits of this book will not permit the rather difficult explanation as to how it is done for latitudes other than 0° or 90°. With the parallels and meridians projected, the map may be drawn. The student should remember that all maps which make any claim to accuracy or correctness are made by locating points of an area to be represented according to their latitudes and longitudes; that is, in reference to parallels and meridians. It will be observed that the orthographic system of projection crowds together areas toward the outside of the map and the scale of miles suitable for the central portion will not be correct for the outer portions. For this reason a scale of miles never appears upon a hemisphere made on this projection.

SUMMARY

In the orthographic projection:

1. The eye is conceived to be at an infinite distance.
2. Meridians and parallels are farther apart toward the center of the map.
3. When a point in the equator is the center, parallels are straight lines.
4. When a pole is at the center, meridians are straight lines. If the northern hemisphere is represented, north is not toward the top of the map but toward the center.

STEREOGRAPHIC PROJECTION

In the stereographic projection the eye is conceived to be upon the surface of the globe, viewing the opposite hemisphere. Points on the opposite hemisphere are projected upon a plane tangent to it. Thus in Figure 68 the eye is at *E* and sees *A* at *A'*, *B* at *B'*, *C* at *C'*, etc. If the earth were transparent, we should see objects on the opposite half of the globe from the view point of this projection.

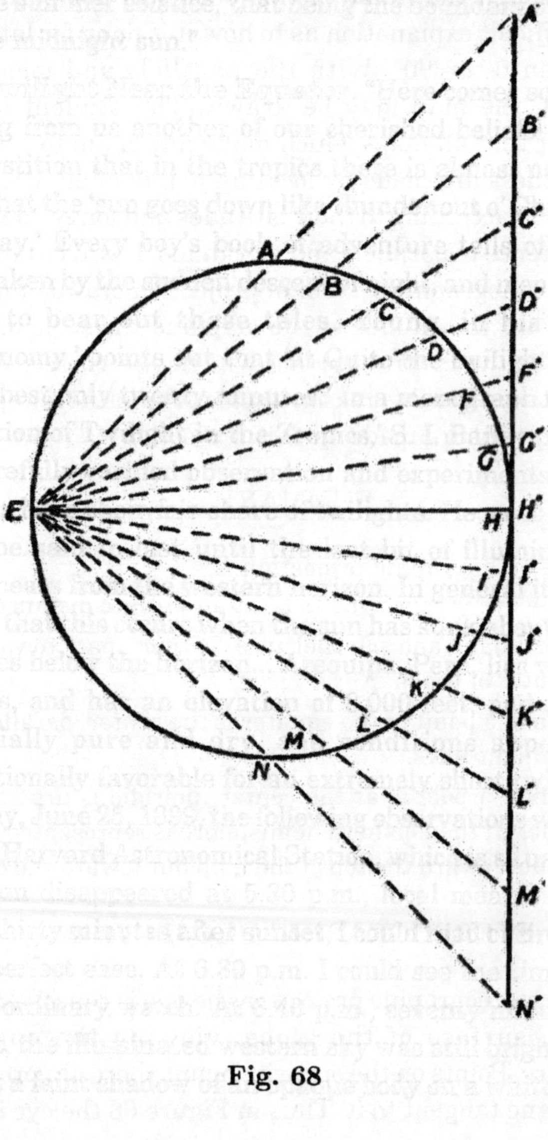

Fig. 68

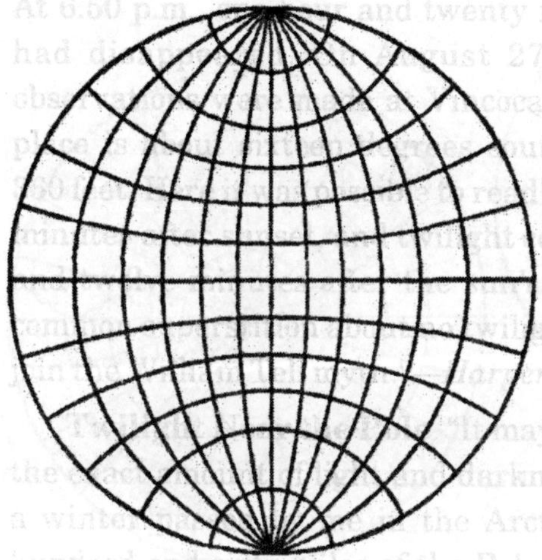

Fig. 69. Equatorial stereographic projection

How to Lay off an Equatorial Stereographic Projection. In Figure 68, *E* represents the eye at the equator, *A* and *N* are the poles and *A′ N′* . is the corresponding meridian of the projection with B′ C′, etc., as the points where the parallels cross the meridian. Taking the line *A′ N′* . of Figure 68 as diameter, Fig. 69. Equatorial stereographic construct upon it a circle projection (see Fig. 69). Divide the circumference into twenty-four equal parts and draw parallels as arcs of circles. Lay off the equator and subdivide it the same as the central meridian, that is, the same as *A′ N′* . of Figure 68. Through the points in the equator, draw meridians as arcs of circles and the projection is complete.

The Polar Stereographic Projection is made on the same plan as the polar orthographic projection, excepting that the parallels have the distances from the pole that are represented by the points in A′ N′ . of Figure 68 (see Figs. 70, 71). Areas are crowded together toward the center of the map when made on the stereographic projection and a scale of miles suitable for the central portion would be too small for the outer portion. This projection is often used, however, because it is so easily laid off.

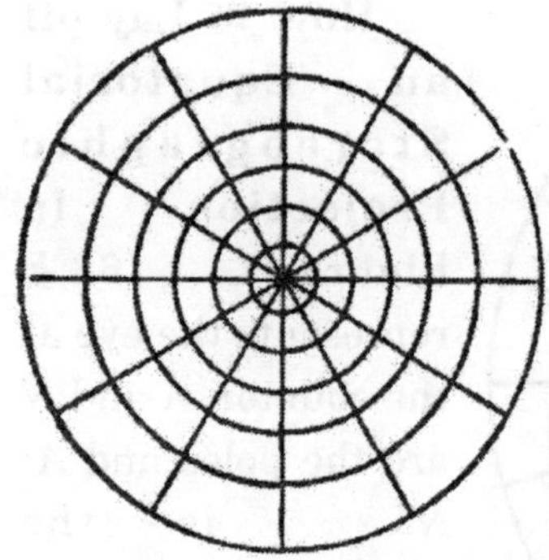

Fig. 70. Polar stereographic projection

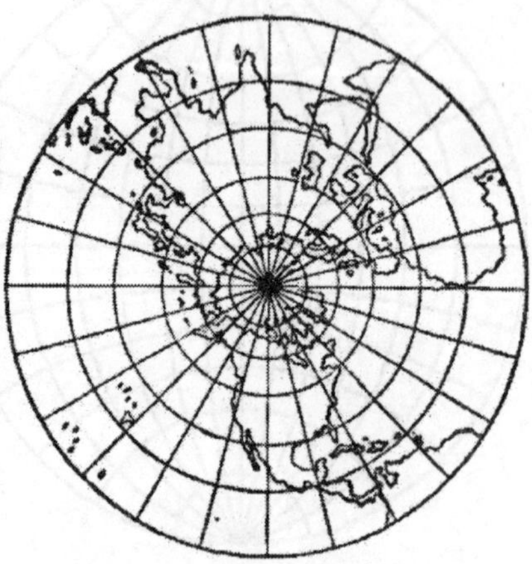

Fig. 71. Northern hemisphere in polar stereographic projection

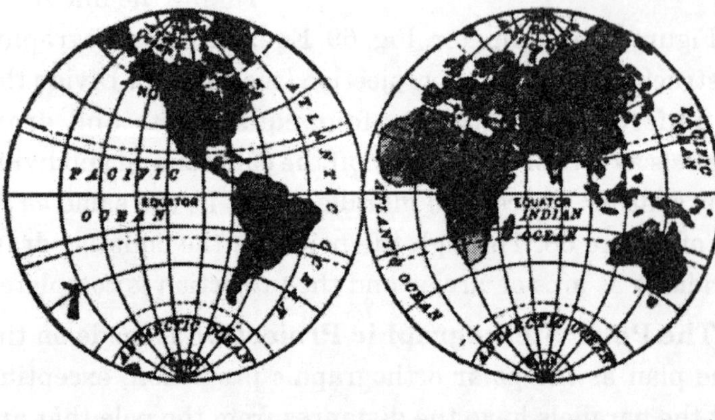

Fig. 72. Hemispheres in equatorial stereographic projection

SUMMARY

In the stereographic projection:

1. The eye is conceived to be on the surface of the globe.

2. Meridians and parallels are nearer together toward the center of the map.
3. When a point in the equator is the center of the map, parallels and meridians are represented as arcs of circles.
4. When a pole is the center, meridians are straight lines.

GLOBULAR PROJECTION.

With the eye at an infinite distance (as in the orthographic projection), parallels and meridians are nearer together toward the *outside* of the map; with the eye on the surface (as in the stereographic projection), they are nearer together toward the *center* of the map. It would seem reasonable to expect that with the eye at some point intermediate between an infinite distance from the surface and the surface itself that the parallels and meridians would be equidistant at different portions of the map. That point is the sine of an angle of 45°, or a little less than the length of a radius away from the surface. To find this distance at which the eye is conceived to be placed in the globular projection, make a circle of the same size as the one which is the basis of the map to be made, draw two radii at an angle of 45° (one eighth of the circle) and draw a line, *AB*, from the extremity of one radius perpendicular to the other radius. The length of this perpendicular is the distance sought (*AB*, Fig. 73).

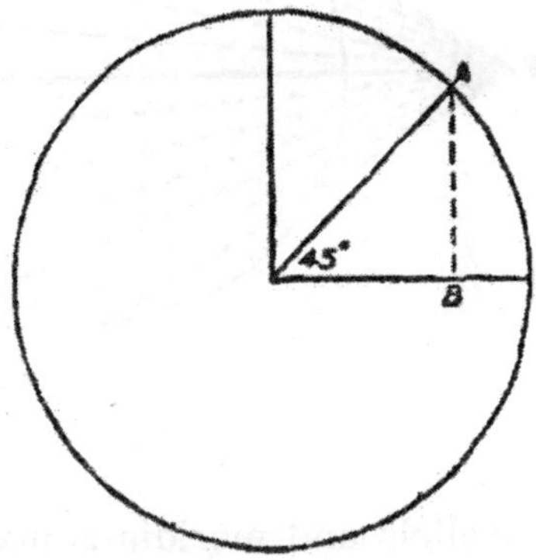

Fig. 73

Thus with the eye at *E* (Fig. 74) the pole *A* is projected to the tangent plane at A', *B* at B', etc., and the distances A'

B´, *B´C´*, etc., are practically equal so that they are constructed as though they were equal in projecting parallels and meridians.

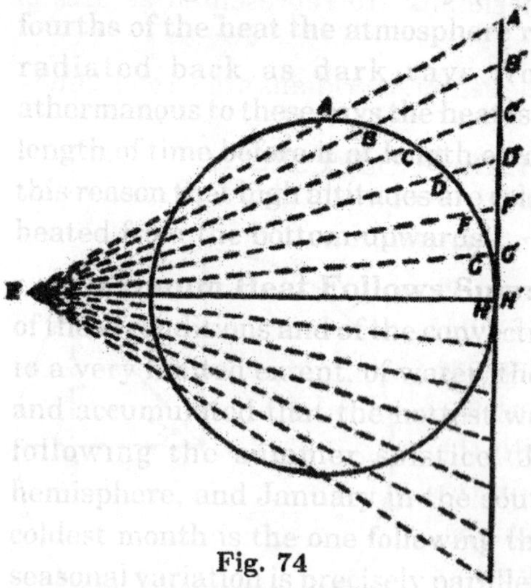

Fig. 74

How to Lay off an Equatorial Globular Projection. As in the orthographic or stereographic projections, a circle is divided into equal parts, according to the number of parallels desired, the central meridian and equator being subdivided into half as many equal parts. Parallels and meridians may be drawn as arcs of circles, being suciently accurate for ordinary purposes (see Fig. 75).

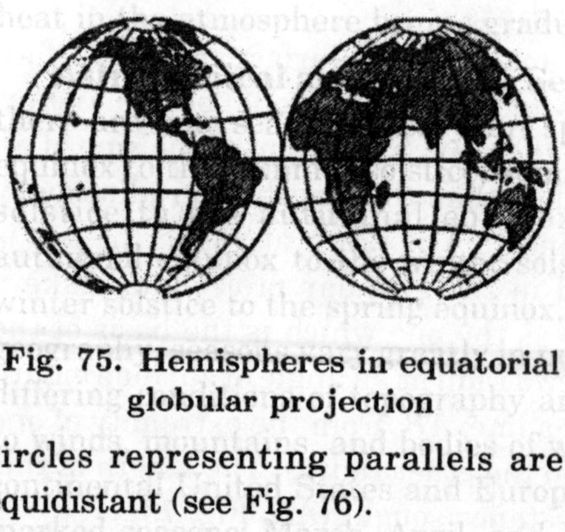

Fig. 75. Hemispheres in equatorial globular projection

The polar globular projection is laid off precisely like the orthographic and the stereographic projections having the pole as the center, excepting that the concentric circles representing parallels are globular projection equidistant (see Fig. 76).

By means of starlike additions to the polar globular projection (see Fig. 77),the entire globe may be represented.

If folded back, the rays of the star would meet at the south pole. It should be noticed that "south" in this projection is in a line directly away from the center; that is, the top of the map is south, the bottom south, and the sides are also south. While portions of the southern hemisphere are thus spread out, proportional areas are well represented, South America and Africa being shown with little distortion of area and outline.

The globular projection is much used to represent hemispheres, or with the star map to represent the entire globe, because the parallels on a meridian or meridians on a parallel are equidistant and show little exaggeration of areas. For this reasonit is sometimes called an equidistant projection, although there are other equidistant projections. It is also called the De la Hire projection from its discoverer (1640–1718).

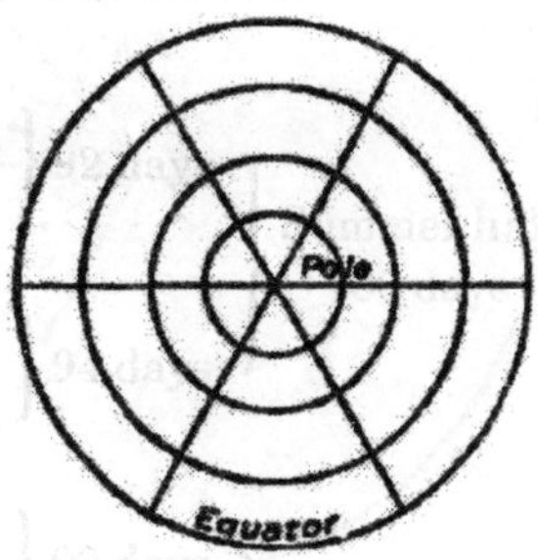

Fig. 76. Polar globular projection

Fig. 77. World in polar globular projection

SUMMARY

In the globular projection:

1. The eye is conceived to be at a certain distance from the globe (sine 45
2. Meridians are divided equidistantly by parallels, and parallels are divided equidistantly by meridians.
3. When a pole is the center of the map, meridians are straight lines.
4. There is little distortion of areas.

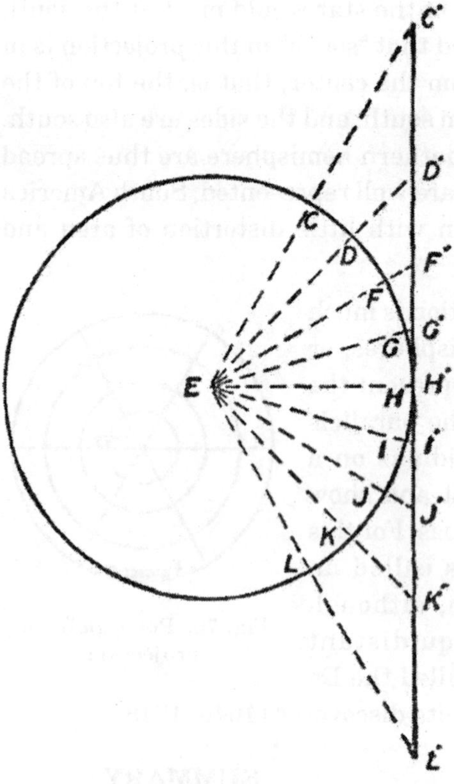

Fig. 78

The Gnomonic Projection When we look up at the sky we see what appears to be a great dome in which the sun, moon, planets, and stars are located. We seem to be at the center of this celestial sphere, and were we to imagine stars and other heavenly bodies to be projected beyond the dome to an imaginary plane we should have a gnomonic projection. Because of its obvious convenience in thus showing the position of celestial bodies, this projection is a very old one, having often been used by the ancients for celestial maps.

Since the eye is at the center for mapping the celestial sphere, it is conceived to be at the center of the earth in projecting parallels and meridians of the earth. As will be seen from Figure 78, the distortion is very great away from the center of the map and an entire hemisphere cannot be shown.

All great circles on this projection are represented as straight lines. This will be apparent if one imagines himself at the center of a transparent globe having parallels and meridians traced upon it. Since the plane of every great circle passes through the center of the globe, the eye at that point will see every portion of a great circle as in one plane and will project it as a straight line. As will be shown later, it is because of this fact that sailors frequently use maps made on this projection.

To Lay off a Polar Gnomonic Projection. Owing to the fact that parallels get so much farther apart away from the center of the map, the gnomonic projection is almost never made with any other point than the pole for center, and then only for latitudes about forty degrees from the pole. The polar gnomonic projection is made like the polar projec-tions previously described,excepting that parallels intersect the meridians at the distancesrepresented in Figure 78.The meridians, being great circles,are represented as straight lines and the parallels as concentriccircles.

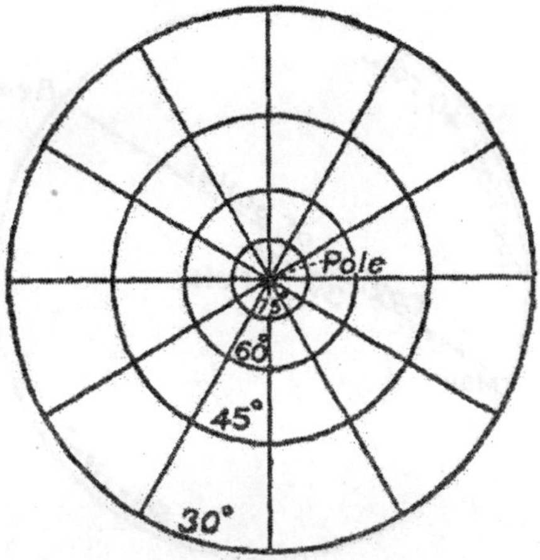

Fig. 79. Polar gnomonic projection

Great Circle Sailing. It would seem at first thought that a ship sailing to a point due eastward, say from New York to Oporto, would follow the course of a parallel, that is, would sail due eastward. This, however, would not be its shortest course. The solution of the following little catch problem in mathematical geography will make clear the reason for this. “A man was forty rods due east of a bear, his gun would carry only thirty rods, yet with no change of position he shot and killed the bear. Where on earth were they?" Solution: This could occur only near the pole where parallels are very small circles. The bear was westward from the man and westward is along the course of a parallel. The bear was thus distant forty rods in a curved line from the man but the bullet ew in a straight line (see Fig. 80).

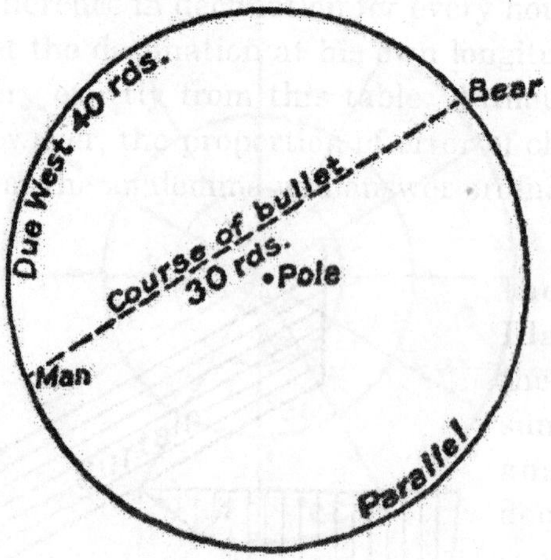

Fig. 80

The shortest distance between two points on the earth is along the arc of agreat circle. A great circle passing through New York and Oporto passes a little to the north of the parallel on which both cities are located. Thus it is that the course of vessels plying between the United States and Europe curves, somewhat to the northward of parallels. This following of a great circle by navigators is called great circle sailing. The equator is a great circle and parallels near it are almost of the same length. In sailing within the tropics, therefore, there is little advantage in departing from the course of a parallel. Besides this, the trade winds and doldrums control the choice of routes in that region and the Mercator projection is always used in sailing there. In higher latitudes the gnomonic projection is commonly used.

Although the gnomonic projection is rarely used excepting by sailors, it is important that students understand the principles underlying its construction since the most important projections yet to be discussed are based upon it.

SUMMARY

In the gnomonic projection:

1. The eye is conceived to be at the center of the earth.
2. There is great distortion of distances away from the center of the map.
3. A hemisphere cannot be shown.
4. All great circles are shown as straight lines.
   a. Therefore it is used largely for great circle sailing.
5. The pole is usually the center of the map.

THE HOMOLOGRAPHIC PROJECTION

The projections thus far discussed will not permit the representation of the entire globe on one map, with the exception of the starlike extension of the polar globular projection. The homolographic projection is a most ingenious device which is used quite extensively to represent the entire globe without distortion of areas. It is a modification of the globular projection.

How to Lay off a Homolographic Projection. First lay off an equatorial globular projection, omitting the parallels. The meridians are semi-ellipses, although those which are no more than 90° from the center meridian may be drawn as arcs of circles.

Having laid off the meridians as in the equatorial globular projection, double the length of the equator, extending it equally in both directions, and subdivide these extensions as the equator was subdivided. Through these points of subdivision and the poles, draw ellipses for meridians.

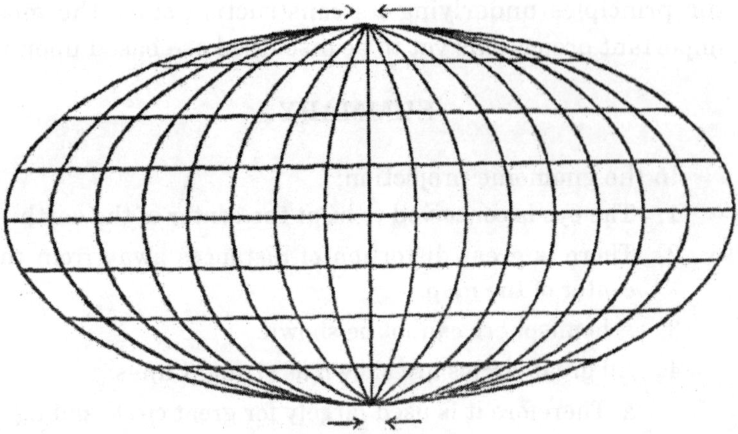

Fig. 81. Homolographic projection

To draw the outer elliptical meridians. Set the points of the compasses at the distance from the point through which the meridian is to be drawn to the central meridian. Place one point of the compasses thus set at a pole and mark off points on the equator for foci of the ellipse. Drive pins in these foci and also one in a pole. Around these three pins form a loop with a string. Withdraw the pin at the pole and draw the ellipse as described on page 22. This process must be repeated for each pair of meridians.

The parallels are straight lines, as in the orthographic projection, somewhat nearer together toward the poles. If nine parallels are drawn on each side of the equator, they may be drawn in the following ratio of distances, beginning at the equator: 2, $1\frac{8}{9}$, $1\frac{7}{9}$, $1\frac{6}{9}$, $1\frac{5}{9}$, $1\frac{4}{9}$, $1\frac{3}{9}$, $1\frac{2}{9}$, $1\frac{1}{9}$. This will give an approximately correct representation.

One of the recent books to make frequent use of this projection is the "Commercial Geography" by Gannett,

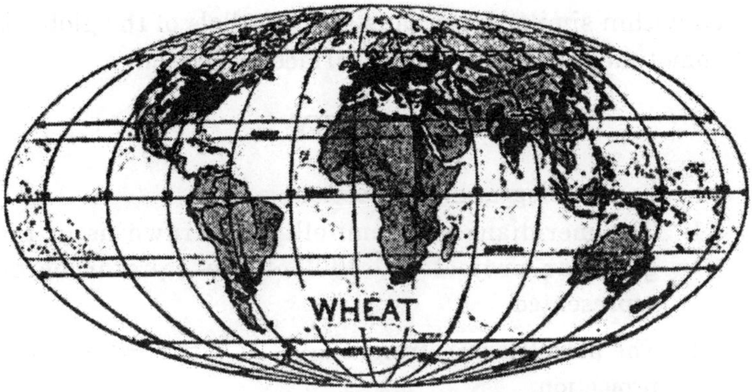

Fig. 82. World in homolographic projection

Garrison, and Houston (see Fig. 82).

Equatorial Distances of Parallels. The following table gives the exact relative distances of parallels from the equator. Thus if a map twenty inches wide is to be drawn, ten inches from equator to pole, the first parallel will be .69 of an inch from the equator, the second 1.37 inches, etc.

ϕ	Distance	ϕ	Distance	ϕ	Distance	ϕ	Distance	ϕ	Distance	ϕ	Distance
5°	.069	20°	.272	35°	.468	50°	.651	65°	.814	80°	.945
10	.137	25	.339	40	.531	55	.708	70	.862	85	.978
15	.205	30	.404	45	.592	60	.762	75	.906	90	1.000

The homolographic projection is sometimes called the Mollweide projection from its inventor (1805), and the Babinet, or Babinet-homolographic projection from a noted cartographer who used it in an atlas (1857). From the fact that within any given section bounded by parallels and meridians, the area of the surface of the map is equal to the area within similar meridians and parallels of the globe, is it sometimes called the equal-surface projection.

SUMMARY

In the homolographic projection:

1. The meridians are semi-ellipses, drawn as in the globular projection, 360° of the equator being represented.
2. The parallels are straight lines as in the orthographic projection.
3. Areas of the map represent equal areas of the globe.
4. There is no distortion of area and not a very serious distortion of form of continents.
5. The globe is represented as though its surface covered half of an exceedingly oblate spheroid.

THE VAN DER GRINTEN PROJECTION

The homolographic projection was invented early in the nineteenth century. At the close of the century Mr. Alphons Van der Grinten of Chicago invented another projection by which the entire surface of the earth may be represented. This ingenious system reduces greatly the angular distortion incident to the homolographic projection and for the inhabitable portions of the globe there is very little exaggeration of areas.

In the Van der Grinten projection the outer boundary is a meridian circle, the central meridian and equator are straight lines, and other parallels and meridians are arcs of circles. The area of the circle is equal to the surface of a globe of one half the diameter of this circle. The equator is divided into 360°, but the meridians are, of course, divided into 180°.

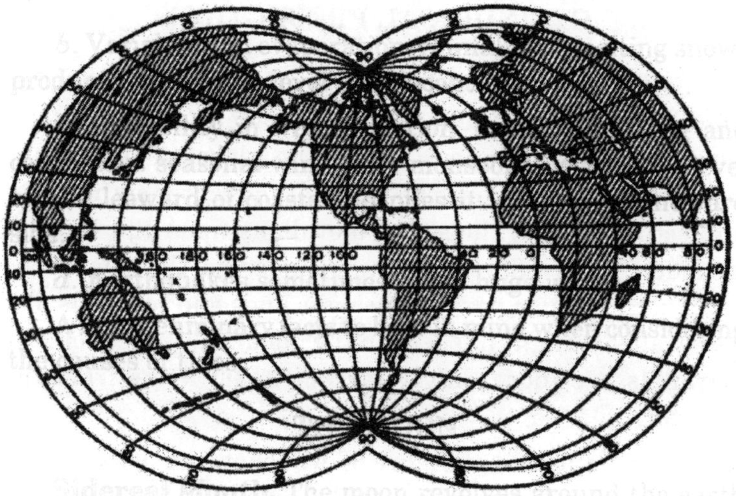

Fig. 83. World in Van der Grinten projection

A modification of this projection is shown in Figure 83. In this the central meridian is only one half the length of the equator, and parallels are at uniform distances along this meridian.

Fig. 84. World in Van der Grinten projection

CYLINDRICAL PROJECTIONS

Gnomonic Cylindrical Projection. In this projection the sheet on which the map is to be made is conceived to be wrapped as a cylinder around the globe, touching the equator. The eye is conceived to be at the center of the globe, projecting the parallels and meridians upon the tangent cylinder. Figure 85 shows the cylinder partly unwrapped with meridians as parallel straight lines and parallels also as parallel straight lines. As in the gnomonic projection, the parallels are increasingly farther apart away from the equator.

An examination of Figure 86 will show the necessity for the increasing distances of parallels in higher latitudes. The eye at the center *(E)* sees *A* at *A´*, *B* at *B´*, etc. Beyond 45° from the equator the distance between parallels becomes very great. *A´ B.* represents the same distance (15° of latitude) as *G´ H´*, but is over twice as long on the map. At *A.* (60° north latitude) the meridians of the globe are only half as far apart as they are at the equator, but they are represented on the map as though they were just as far apart there as at the equator. Because of the rapidly increasing distances of

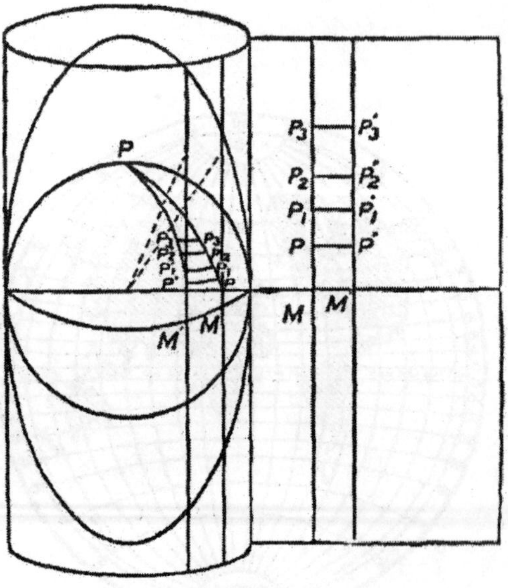

Fig. 85

parallels, to represent higher latitude than 60°. would require a very large sheet, so the projection is usually modified for a map of the earth as a whole, sometimes arbitrarily.

G′ H′ is the distance from the equator to the first parallel, and since a degree of latitude is about equal to a degree of longitude there, this distance may be taken between meridians.

Stereographic Cylindrical Projection. For reasons just given, the gnomonic or central cylindrical projection needs reduction to show the poles at all or any high latitudes without great distortion. Such a reduction is well shown in the stereographic projection. In this the eye is conceived to be on the equator, projecting each meridian from the view point of the meridian opposite to it. Figure 87 shows the plan on which it is laid o", meridians being parallel straight lines and equidistant and parallels being parallel straight lines at increasing distances away from the equator.

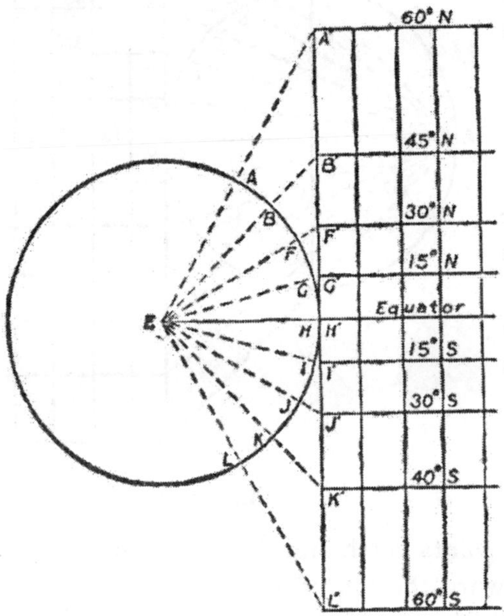

Fig. 86

The Mercator (Cylindrical) Projection. In the orthographic, stereographic, globular, gnomonic, homolographic, and Van der Grinten projections, parallels or meridians, or both, are represented as curved lines. It should be borne in mind that directions on the earth are determined from parallels and meridians. North and south are along a meridian and when a meridian is represented as a curved line, north and south are along that curved line. Thus the two arrows shown at the top of Figure 81, are pointing in almost exactly opposite directions and yet each is pointing due north. The arrows at the bottom point opposite each other, yet both point due south. The arrows pointing to the right point the same way, yet one points north and the other points south. A line pointing toward the top of a map may or may not point north. Similarly, parallels lie in a due east-west direction and to the right on a map may or may not be to the east.

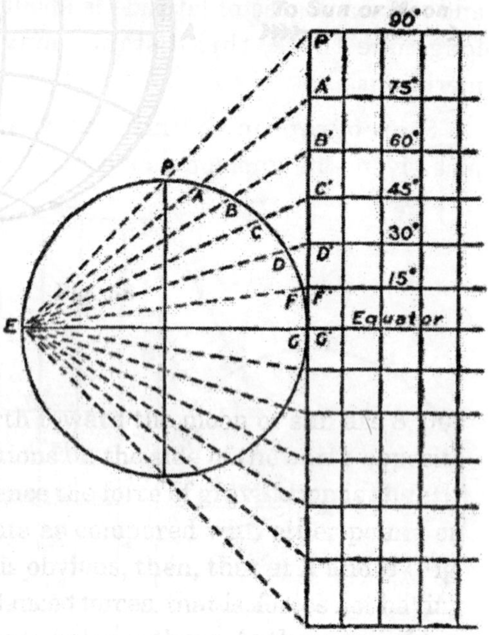

Fig. 87

It should be obvious by this time that the map projections studied thus far represent directions in a most unsatisfactory manner, however well they may represent areas. Now to the sailor the principal value of a chart is to show directions to steer his course by and if the direction is represented by a curved line it is a slow and difficult process for him to determine his course. We have seen that the gnomonic projection employs straight lines to represent arcs of great circles, and, consequently, this projection is used in great circle sailing. The Mercator projection shows all parallels and meridians as straight lines at proportional distances, hence directions as straight lines, and is another, and the only other, kind of map used by sailors in plotting their courses.

Maps in Ancient Times. Before the middle of the fifteenth century, sailors did not cover very great portions of the earth's surface in continuous journeys out of sight of land where they had to be guided almost wholly by the stars. Mathematical accuracy in maps was not of very great importance in navigation until long journeys had to be made with no opportunity for verification of calculations. Various roughly accurate map projections were made. The map sent to Columbus about the year 1474 by the Italian astronomer Toscanelli, with which he suggested sailing directions across the "Sea of Darkness," is an interesting illustration of a common type of his day.

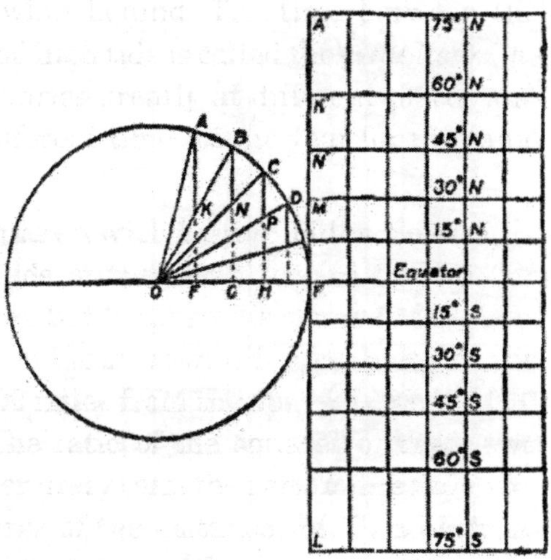

Fig. 89

The long journeys of the Portuguese along the coast of Africa and around to Asia and the many voyages across the

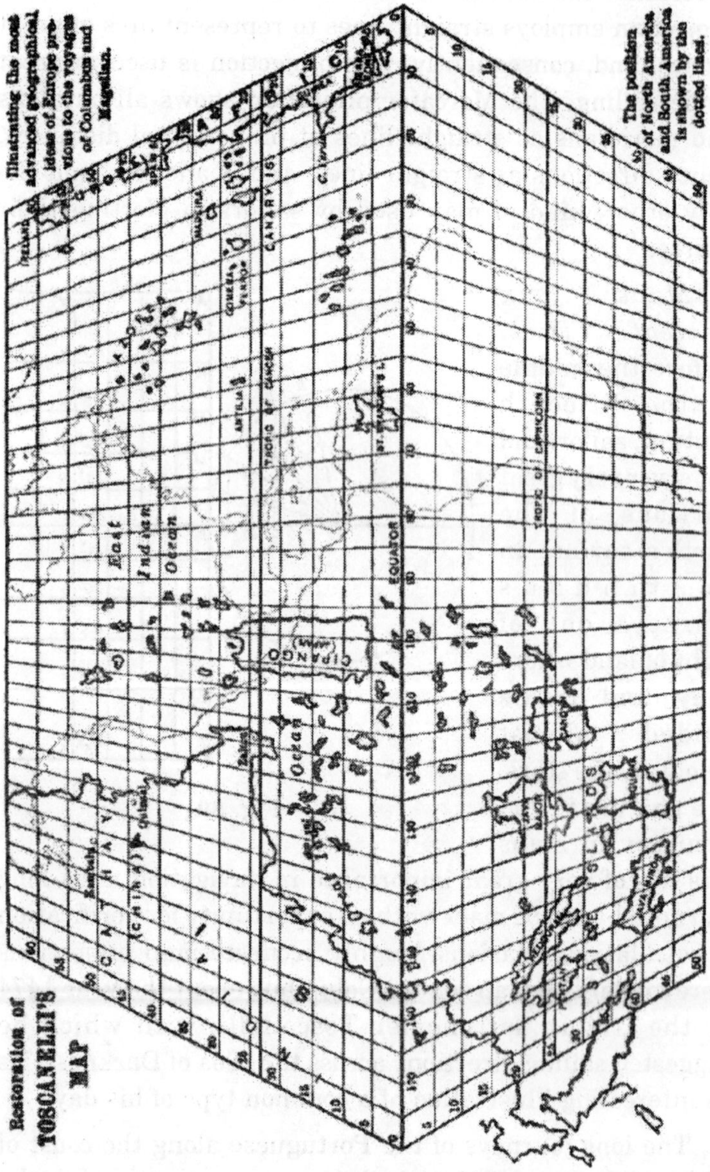
Restoration of
TOSCANELLI'S
MAP
Illustrating the most advanced geographical ideas of Europe previous to the voyages of Columbus and Magellan.
The position of North America and South America is shown by the dotted lines.
CIPANGO
EQUATOR
TROPIC OF CANCER
TROPIC OF CAPRICORN

Fig. 88

Atlantic early in the sixteenth century, made accurate map projection necessary. About the middle of that century, Emperor Charles V of Spain employed a Flemish mathematician named Gerhard Kramer to make maps for the use of his sailors. The word Kramer means, in German, "retail merchant," and this translated into Latin, then the universal language of science, becomes Mercator, and his invention of a very valuable and now widely used map projection acquired his Latinized name.

Plan of Mercator Chart. The Mercator projection is made on the same plan as the other cylindrical projections, excepting as to the distances between parallels. The meridians are represented as parallel lines, whereas on the globe they conThere is thus a distortion of longitudes, greater and greater, away from the equator. Now the Mercator projection makes the parallels farther apart away from the equator, exactly proportional to the meridional error. Thus at latitude 60° the meridians on the earth are almost exactly half as far apart as at the equator, but being equidistant on the map, they are represented as twice as far apart as they should be. The parallels in that portion of the Mercator map are accordingly made twice as far apart as they are near the equator. Since the distortion in latitude exactly equals the distortion in longitude and parallels and meridians are straight lines, all directions are represented as straight lines. A navigator has simply to draw upon the map a line from the point where he is to the point to which he wishes to sail in a direct course, measure the angle which this line forms with a parallel or meridian, and steer his ship according to the bearings thus obtained.

To Lay off a Mercator Projection. Figure 89 shows the simplest method of laying off this projection. From the extremity of each radius drop a line to the nearest radius, parallel to the tangent *A′ L*. The lengths of these lines, respectively, represent the distances* between parallels. Thus $N'M$ equals CP, $K'N'$ equals BN, $A'K'$ equals AK The meridians are equidistant and are the same distance apart as the first parallel is from the equator.

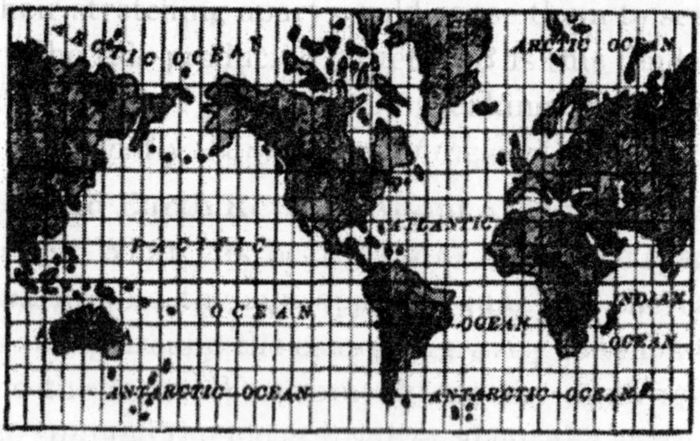

Fig. 90. World in mercator projection

The table of meridional parts on page 217 gives the relative distances of parallels from the equator. By means of this table a more exact projection may be laid off than by the method just suggested. To illustrate: Suppose we wish a map about twenty inches wide to include the 70th parallels. We find in the table that 5944.3 is the distance to the equator. Then, since the map is to extend 10 inches on each side of the equator, $\frac{10}{5944.3}$ is the scale to be used in making the map; that is 1 inch on the map will be represented by 10 inches ÷ 5944.3. Suppose we wish to lay off parallels ten degrees apart. The first parallel to be drawn north of the equator has, according to the table, 599.1 for its meridional distance. This multiplied by $\frac{10}{5944.3}$ equals slightly more than 1. Hence

*Technically speaking, the distance is the tangent of the angle of latitude and any table of natural tangents will answer nearly as well as the table of meridional parts, although the latter is more accurate, being corrected for the oblateness of the meridian.

the parallel 10° should be laid off 1 inch from the equator. The 20th parallel has for its meridional distance 1217.3. This multiplied by the scale $\frac{10}{5944.3}$ gives 2.03 inches from the equator. The 30th parallel has a meridional distance 1876.9, this multiplied by the scale gives 3.15 inches. In like manner the other parallels are laid o”. The meridians will be $\frac{10}{5944.3}$ × 60 or 600 inches ÷ 5944.3 for every degree, or for ten degrees 6000 inches ÷ 5944.3, which equals 1.01 inches. This makes the map 36.36 inches long (1.01 inches□36 = 36.36 inches).

TABLE OF MERIDIONAL PARTS*

1°	59.6	18°	1091.1	35°	2231.1	52°	3647.1	69°	5773.1
2°	119.2	19°	1154.0	36°	2304.5	53°	3745.4	70°	5944.3
3°	178.9	20°	1217.3	37°	2378.8	54°	3846.1	71°	6124.0
4°	238.6	21°	1281.0	38°	2454.1	55°	3949.1	72°	6313.0
5°	298.4	22°	1345.1	39°	2530.5	56°	4054.9	73°	6512.4
6°	358.3	23°	1409.7	40°	2607.9	57°	4163.4	74°	6723.6
7°	418.3	24°	1474.7	41°	2686.5	58°	4274.8	75°	6948.1
8°	478.4	25°	1540.3	42°	2766.3	59°	4389.4	76°	7187.8
9°	538.6	26°	1606.4	43°	2847.4	60°	4507.5	77°	7444.8
10°	599.1	27°	1673.1	44°	2929.9	61°	4628.1	78°	7722.1
11°	659.7	28°	1740.4	45°	3013.7	62°	4754.7	79°	8023.1
12°	720.6	29°	1808.3	46°	3099.0	63°	4884.5	80°	8352.6
13°	781.6	30°	1876.9	47°	3185.9	64°	5018.8	81°	8716.4
14°	842.9	31°	1946.2	48°	3274.5	65°	5158.0	82°	9122.7
15°	904.5	32°	2016.2	49°	3364.7	66°	5302.5	83°	9583.0
16°	966.4	33°	2087.0	50°	3456.9	67°	5452.8	84°	10114.0
17°	1028.6	34°	2158.6	51°	3551.0	68°	5609.5	85°	10741.7

We see, then, that the same scale of miles cannot be used for different parts of the map, though within 30° of the equator representations of areas will be in very nearly true proportions. The parallels in a map not wider than this, say

*From Bowditch's *Practical Navigator*.

for Africa, may be drawn equidistant and the same distance apart as the meridians, the inaccuracy not being very great.

SUMMARY

In the cylindrical projection:

1. A cylinder is conceived to be wrapped around the globe, tangent to the equator.
2. All parallels and meridians are represented as straight lines, the former intersecting the latter at right angles.
3. The parallels are made at increasing distances away from the equator:

   a. In the gnomonic projection, as though projected from the center of the earth to the tangent cylinder.

   b. In the stereographic projection, as projected from the equator upon an opposite meridian, the projection point varying for each meridian.

   c. In the Mercator projection, at distances proportional to the meridional excess.

   Directions are better represented in this projection than in any other. Here northward is directly toward the top of the map, eastward directly toward the right, etc. For this reason it is the projection most commonly employed for navigators' charts.
4. There is great distortion of areas and outlines of continents in high latitudes; Greenland appears larger than South America.
5. The entire globe may be represented in one continuous map.
6. The same scale of miles cannot be used for high latitudes that is used near the equator.

CONIC PROJECTION

The portion of a sphere between the planes of two parallels which are near together is very similar to the zone of a cone (see Fig. 91). Hence, if we imagine a paper in the form of a cone placed upon the globe and parallels and meridians projected upon this cone from the center of the globe, then this conical map unrolled, we can understand this system.

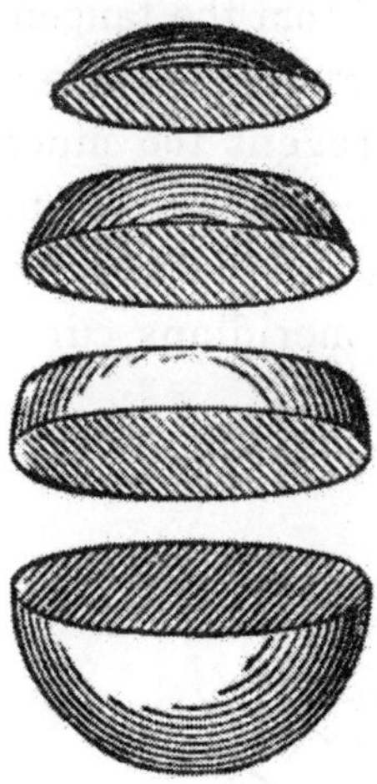

Fig. 91

Along the parallel tangent to the cone, points on the map will correspond exactly to points upon the globe. Parallels which are near the line of tangency will be represented very much in the relative positions they occupy on the globe. In a narrow zone, therefore, near the tangent parallel, there will be very little distortion in latitudes and longitudes and an area mapped within the zone will be very similar in form and area to the form and area as it appears upon the globe itself. For this reason the conic projection, or some modification of it, is almost always employed in representing small areas of the earth's surface.

To Lay off a Conic Projection. If the forty-fifth parallel is the center of the area to be mapped, draw two straight lines tangent to the forty-fifth parallel of a circle (see Fig. 93). Project upon these lines points for parallels as in the gnomonic projection. With the apex as center, draw arcs of circles through these points for parallels. Meridians are straight lines meeting at the apex and are equidistant along any parallel.

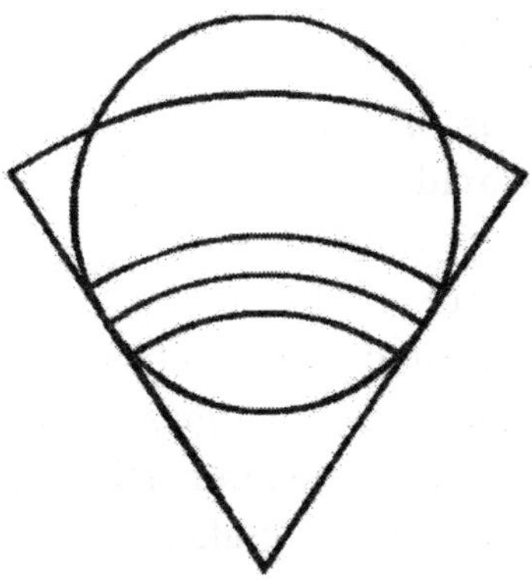

Fig. 92

It will be observed that parallels are farther apart away from the tangent parallel (45°, in this case) as in the Mercator projection they are farther apart away from the equator, which is tangent to the globe in that projection. There is also an exaggeration of longitudes away from the tangent parallel. Because of this lengthening of parallels, meridians are sometimes curved inwardly to prevent too much distortion of areas. The need for this will be apparent if one draws parallels beyond the equator, for he will find they are longer than the equator itself unless meridians curve inwardly there.

By taking the tangent parallel ten degrees north of the equator and reducing distances of parallels, a fan-shaped map of the world may be shown. In this map of the world on the conic projection, there is even greater distortion of parallels south of the equator, but since meridiansconverge somewhat north of the equator there is less distortion in northern latitudes. Since most of the land area of the globe is in the northern hemisphere, this projection is much better suited to represent the entire world than the Mercator projection.

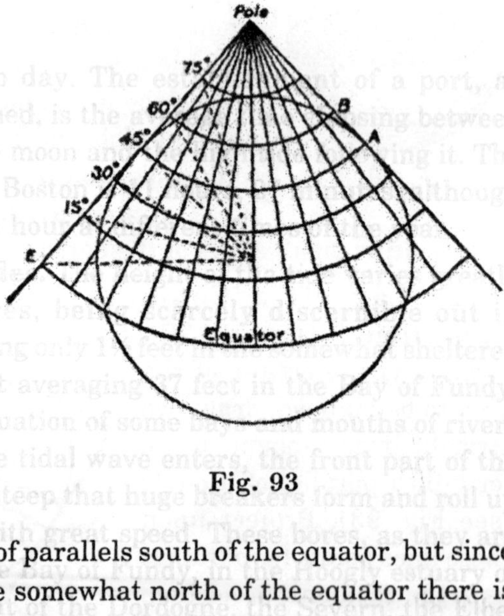

Fig. 93

Fig. 94. North America in conic projection

Bonne's (Conic) Projection. This is a modification of the conic projection as previously described to prevent exaggeration of areas away from the parallel which is conceived to be touching the globe. The central meridian is a straight line and parallels are concentric equidistant circles. The distance between parallels is the length of the arc of the circle which is used as a basis for the projection. For ordinary purposes the distance *AB* (Fig. 93) may be taken for each of the distances between parallels.

Having laid off the central meridian and marked off the arcs for parallels, the true distance of the meridian on each parallel is laid off and the meridian is drawn through these points. This gives a gentle inward curve for meridians toward the outside of the map of continents. Instead of following Bonne's

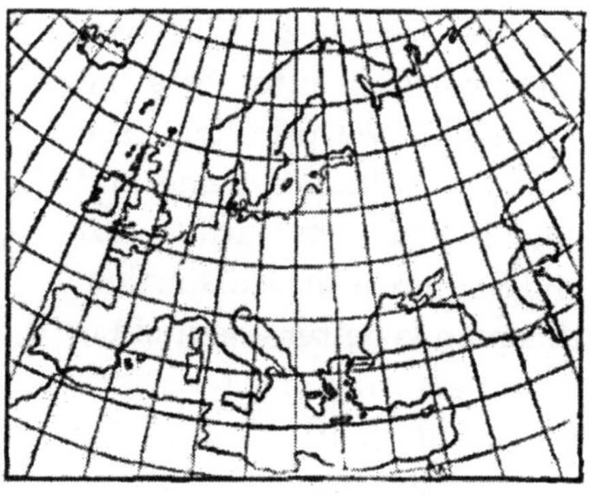

Fig. 96. Europe in conic projection

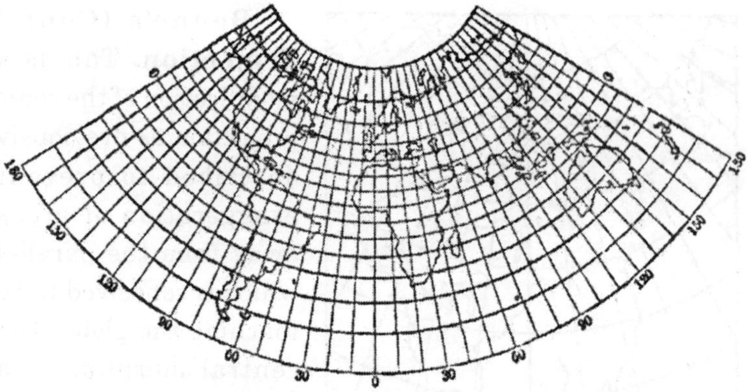

Fig. 95. The world in conic projection

system with strict accuracy, the map maker sometimes makes the curve a little less in lower latitudes, allowing a slight exaggeration of areas to permit the putting in of more details where they are needed.

Intersecting Conic Projection. Where a considerable extent in latitude is to be represented, the cone is sometimes conceived to cut into the sphere. In this case, each meridian intersects the sphere at two parallels (see Fig. 97) and since along and near the tangent parallels *(A* and *B)* there is little distortion, this plan is better adapted for a map showing greater width north and south than is the conic projection.

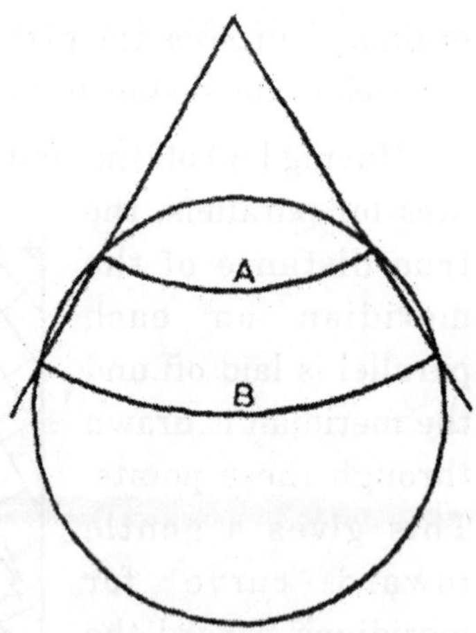

Fig. 97

The map of Europe well illustrates this difference. Europe

lies between 35° and 75° north latitude. On a conic projection the tangent parallel would be 55° Near this parallel there would be no exaggeration of areas but at the extreme north and south 20° away from this parallel, there would be considerable distortion. If, instead, we make an intersecting conic projection, we should have the cone pass through parallels 45° and 65° and along these parallels there would be no distortion and no part of the map being more than 10° away from these lines, there would be very little exaggeration anywhere.

Fig. 98. Africa and Europe in polyconic projection

It should be noticed that the region between the intersections of the meridians must be projected back toward the center of the sphere and thus be made smaller in the map than it appears on the globe. The central parallel would be too short in proportion to the rest. Since this area of Europe (between 45° and 65°) is the most important portion and should show most details, it would be a serious defect, from the practical map maker's point of view, to minify it.

Polyconic Projection. This differs from the conic projection in that it is readjusted at each parallel which is drawn, so that each one is tangent to the sphere. This makes the circumscribing cone bent at each parallel, a series of conic sections. The word polyconic means "many cones." The map constructed on this projection is thus accurate along each parallel, instead of along but one as in the conic projection or along two as in the intersecting conic projection. For representing small areas this is decidedly the most accurate projection known. Since the zone along each parallel is projected on an independent cone, the point which is the apex for one cone will not be the same for any other (unless both north and south latitudes are shown in the same map). In the conic projection the parallels are all made from the apex of the cone as the center. In the polyconic projection each parallel has its own conical apex and hence its own center. This may easily be observed by a comparison of the parallels in Figure 94 (conic projection, all made from one center) and those in Figure 98 (polyconic projection, each made from a different center).

SUMMARY

In the conic projection:

1. A cone is conceived to be fitted about a portion of the globe, tangent to some parallel.
2. The tangent parallel shows no distortion and portions near it have but little. This projection is therefore used extensively for mapping small areas.

a. In the conic projection on the gnomonic or central plan, the eye is conceived to be at the center of the globe, parallels are crowded closer together toward the central parallel, and distant areas are exaggerated.

The cone may be conceived to intersect the globe at two parallels, between which there is a diminution of areas and beyond which there is an exaggeration of areas.

b. In the Bonne projection parallels are drawn at equidistant intervals from a common center and meridians are slightly curved to prevent distortion in longitudes.

c. In the polyconic projection many short conic sections are conceived to be placed about the globe, one for each parallel represented. Parallels are drawn from the apexes of the cones.

THE SCALE

The area of any map bears some proportion to the actual area represented. If the map is so drawn that each mile shall be represented by one inch on the map, since one mile equals 63,360 inches, the scale is said to be 1 : 63,360. This is often written fractionally, $\frac{1}{63,369}$. A scale of two inches to the mile is 1 : 31,680. These, of course, can be used only when small areas are mapped. The following scales with their equivalents are most commonly used in the United States Geological Survey, the first being the scale employed in the valuable geological folios covering a large portion of the United States.

Scale 1 : 125, 000, 1 mile = 0.50688 inches.
Scale 1 : 90, 000, 1 mile = 0.70400 inches.
Scale 1 : 62, 500, 1 mile = 1.01376 inches.
Scale 1 : 45, 000, 1 mile = 1.40800 inches.

SOME CONCLUSIONS

The following generalizations from the discussion of map projections seem appropriate.

1. In all maps north and south lie along meridians and east and west along parallels. The top of the map may or may not be north; indeed, the cylindrical projection is the only one that represents meridians by perpendicular lines.

2. Maps of the same country on different projections may show different shapes and yet each may be correct. To make maps based upon some arbitrary system of triangles or lines is not scientific and often is not even helpful.

3. Owing to necessary distortions in projecting the parallels and meridians, a scale of miles can rarely be used with accuracy on a map showing a large area.

4. Straight lines on maps are not always the shortest distances between two points. This will be clear if we remember that the shortest distance between two points on the globe is along the arc of a great circle. Now great circles, such as meridians and the equator, are very often represented as curved lines on a map, yet along such a curved line is the shortest distance between any two places in the line on the globe which the map represents.

5. To ascertain the scale of miles per inch used on any map, or verify the scale if given, measure the space along a meridian for one inch and ascertain as correctly as possible the number of degrees of latitude contained in the inch. Multiply this by the number of miles in one degree of latitude, 69, and you have the number of miles on the earth represented by one inch on the map.

CHAPTER XI

THE UNITED STATES GOVERNMENT LAND SURVEY

Allowance for Curvature. One of the best proofs that the earth is a sphere is the fact that in all careful measurements over any considerable area, allowance must be made for the curvature of the surface. If two lines be drawn due northward for one mile in the northern part of the United States or in central Europe, say from the 48th parallel, they will be found nearer together at the northern extremities than they are at the southern ends.

Origin of Geometry. One of the greatest of the practical problems of mathematics and astronomy has been the systematic location of lines and points and the measurement of surfaces of the earth by something more definite, more easily described and relocated than metes and bounds. Indeed, geometry is believed to have had its origin in the need of the ancient Egyptians for surveying and relocating the boundaries of their lands after the Nile oods.

Locating by Metes and Bounds. The system of locating lands by metes and bounds prevails extensively over the world and, naturally enough, was followed in this country by the early settlers from Europe. To locate an area by landmarks, some point of beginning is established and the boundary lines are described by means of natural objects such as streams, trees, well established highways, and stakes, piles of stone, etc., are placed for the purpose. The directions are usually indicated by reference to the magnetic compass and distances as ascertainedby surveyors' chains. But landmarks decay and change, and rivers change their courses.* The magnetic needle of the compass does not point due north (excepting along two or three isogonal lines, called agones), and varies from year to year. This gives rise to endless confusion, uncertainty, and litigation.

36	30	24	18	12	6
35	29	23	17	11	5
34	28	22	16	10	4
33	27	21	15	9	3
32	26	20	14	8	2
31	25	19	13	7	1

Fig. 99

Variation almost without limit occurs in such descriptions, and farms assume innumerable forms, sometimes having a score of angles. The transitory character of such platting of land is illustrated in the following excerpt from a deed to a piece of property in Massachusetts Bay Colony, bearing the date: "Anno Domini one thousand seven hundred and thirty-six

*Where a meandering river constitutes the boundary of a nation or state, changes in the course of the stream give rise to problems in civil government, as the following incident illustrates. A minister in the southern part of South Dakota was called upon to officiate at a wedding in a home in a bend of the Missouri River. During the high water of the preceding spring, the river had burst over the narrow neck at the bend and at the time of the wedding it was owing on both sides of the cut-off so that there was a doubt as to whether the main channel of the stream, the interstate boundary line, was north of them and they were in Nebraska, or south and they were still in South Dakota. To be assured of the legality of the marriage rite, the bridal couple, minister, and witnesses rowed to the north bank, and up on the South Dakota bluff the marriage service was performed, the bridal party returning—they cared not to which state, for the festivities.

and in the tenth year of the reign of our sovereign Lord George the Second, King." In this, Emma Blowers deeds to William Stanley, "A certain parcel of Upland and Swamp Ground Situate and lying in the Township of Manchester being the thirty-first lot into the Westerly Division of Common Rights made in said Manchester by the proprietors thereof in the year of our Lord one thousand six hundred ninety-nine, Said lot containing Ten Acres, more or less, being cutted and bounded as followeth Viz: At the Northeast Corner with a maple tree between Sowest and Abraham Master's, from that Southeasterly thirty-nine poles to Morgan's Stump, so called, from that Southeasterly fourty-four poles upon said west Farm Line to a black Oak tree, from that Sixty-six poles Northward to the first bounds, or however Otherwise the Said Lot is or ought to have been bounded."

Survey of Northwest Territory. When, in 1785, practically all of the territory north and west of the Ohio River had been ceded to the United States by the withdrawal of state claims, Congress provided for its survey, profiting from the experiences resulting from hastily marked boundaries. Thomas Hutchins was appointed Geographer of the United States, and after the selection of thirteen assistants, he was instructed to begin its survey. Starting in 1786 from the southwest corner of Pennsylvania, he laid off a line due north to a point on the north bank of the Ohio River. From this

6	5	4	3	2	1
7	8	9	10	11	12
18	17	16	15	14	13
19	20	21	22	23	24
30	29	28	27	26	25
31	32	33	34	35	36

Fig. 100

point he started a line westward. According to the directions of Congress, every six miles along this east-west "geographer's line," meridians were to be laid off and parallels to it at intervals of six miles, each of the six miles square to be divided into thirty-six square miles and these divided into "quarters," thus spreading a huge "gridiron" over the land. The larger squares were called "townships," an adaptation of the New England "town." They are commonly called "Congressional townships" in most parts of the United States, to distinguish them from the political subdivision of the county called the "civil township" or the "municipal township."

31	32	33	34	35	36
30	29	28	27	26	25
19	20	21	22	23	24
18	17	16	15	14	13
7	8	9	10	11	12
6	5	4	3	2	1

Fig. 101

Jefferson is believed to have suggested this general plan which with some variations has been continued over the major portion of the United States and the western portion of Canada. Hutchins and his crew laid off the "ge-ographer's line" only forty-two miles, making seven ranges of townships west of the Pennsylvania state boundary, when they were frightened away by the Indians. The work was continued, however, on the same general plan one exception being the method of numbering the sections. In these first "seven ranges" the sections are numbered as in Figure 99, elsewhere in the United States they are numbered as in Figure 100, and in western Canada as in Figure 101. Each of the square miles is commonly called a "section."

The law passed by Congress May 20, 1785, provided that,"The surveyors . . . shall proceed to divide the said territory into townships of six miles square, by lines running due north and south, and others crossing these at right angles, as near as may be." Owing to the convergence of the meridians this, of course, was a mathematical impossibility; "as near as may be," however, has been broadly interpreted. According to the provisions of this act and the acts of May 18, 1796, May 10, 1800, and Feb. 11, 1805, and to rules of commissioners of the general land office, a complete system has been evolved, the main features of hich are as follows:

Principal Meridians. These are run due north, south, or north and south from some initial point selected with great care and located in latitude and longitude by astronomical means. Thirty-two or more of these principal meridians have been surveyed at irregular intervals and of varying lengths. Some of these are known by numbers and some by names. The first principal meridian is the boundary line between Indiana and Ohio; the second is west of the center of Indiana, extending the entire length of the state; the third is in the center of Illinois, extending the entire length of the state; the Tallahassee principal meridian passes directly through that city and is only about twenty-three miles long; other principal meridians are named Black Hills, New Mexico, Indian, Louisiana, Mount Diablo, San Bernardino,* etc. To the east, west, or east and west of principal meridians, north and south rows of townships called ranges are laid o". Each principal meridian, together with the system of townships based upon it, is independent of every other principal

*The entire platting of the portions of the United States to which this discussion refers is clearly shown on the large and excellent maps of the United States, published by the Government and obtainable, at the actual cost, eighty cents, from the Commissioner of the General Land offi ce, Washington, D. C.

meridian and where two systems come together irregularities are found.

Base Lines. Through the initial point selected from which to run the principal meridian, an east-west base line is run, at right angles to it, and corresponds to a true geographic parallel. As in case of the principal meridian, this line is laid off with great care since the accuracy of these controlling lines determines the accuracy of the measurements based upon them.

Tiers of townships are laid off and numbered north and south of these base lines. In locating a township the word tier is usually omitted; township number 4 north, range 2 west of the Michigan principal meridian, means the township in tier 4 north of the base line and in the second range west of the Michigan principal meridian. This is the township in which Lansing, Michigan, is located.

The fourth principal meridian in western Illinois and Wisconsin has two base lines, one at its southern extremity extending westward to the Mississippi River and the other constituting the interstate boundary line between Wisconsin and Illinois. The townships of western Illinois are numbered from the southern base line, and all of those in Wisconsin and northeastern Minnesota are numbered from the northern base line. The fourth principal meridian is in three sections, being divided by an eastern bend of the Mississippi River and by the western portion of Lake Superior.

The largest area embraced within one system is that based upon the fifth principal meridian. This meridian extends northward from the mouth of the Arkansas River until it again intersects the Mississippi River in northeastern Missouri and then again it appears in the big eastern bend of the Mississippi River in eastern Iowa. Its base line passes a few miles south of Little Rock, Arkansas, from which fact it is sometimes called the Little Rock base line. From this

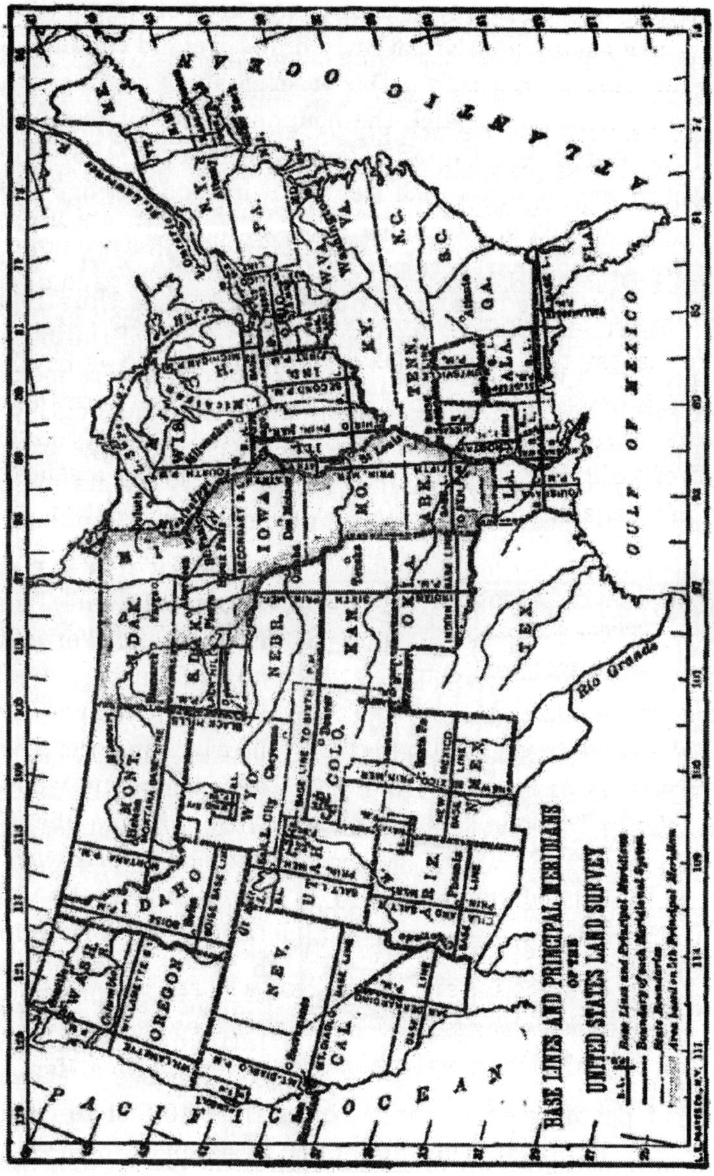
BASE LINES AND PRINCIPAL MERIDIANS
OF THE
UNITED STATES LAND SURVEY
ATLANTIC OCEAN
GULF OF MEXICO
PACIFIC OCEAN
Rio Grande
WASH.
OREGON
CAL.
NEV.
IDAHO
MONT.
WYO.
UTAH
ARIZ.
COLO.
NEBR.
KAN.
OKLA.
TEX.
IOWA
MO.
ARK.
LA.
ILL.
IND.
KY.
TENN.
ALA.
GA.
FLA.
S.C.
N.C.
VA.
W.VA.
PA.
N.Y.

Fig. 102.

meridian and base line all of Arkansas, Missouri, Iowa, North Dakota, and the major portions of Minnesota and South Dakota have been surveyed, an area considerably larger than that of Germany and Great Britain and Ireland combined. The most northern tier from this base lies about a mile south of the forty-ninth parallel, the boundary line between the United States and Canada, and is numbered 163. The southern row of sections of tier 164 with odd lottings lies between tier 163 and Canada. Its most northern township is in the extreme northern portion of Minnesota, west of the Lake of the Woods, and is numbered 168. It thus lies somewhat more than a thousand miles north of the base from which it was surveyed. There are nineteen tiers south of the base line in Arkansas, making the extreme length of this area about 1122 miles. The most eastern range from the fifth principal meridian is numbered 17and its most western, 104, making an extent in longitude of 726 miles.

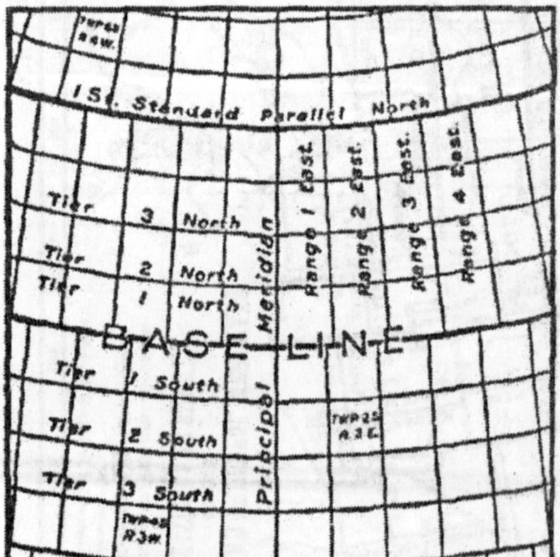

Fig. 103.

Standard Parallels. The eastern and western boundaries of townships are, as nearly as may be, true meridians, and when they have been extended northward through several tiers, their convergence becomes considerable. At latitude 40° the convergence is about 6.7 feet per mile or somewhat more than 40 feet to each township. To prevent this diminution in size of townships to the north of the base line, standard parallels are run, along which six-mile measurements are made for a new set of townships. These lines are also called *correction lines* for obvious reasons.

Division of Dakotas. When Dakota Territory was divided and permitted to enter the Union as two states, the dividing line agreed upon was the *seventh standard parallel* from the base line of the fifth principal meridian. This line is about four miles south of the parallel 46° from the equator and was chosen in preference to the geographic parallel because it was the boundary line between farms, sections, townships, and, to a considerable extent, counties. The boundary line between Minnesota and Iowa is what is called a secondary base line and corresponds to a standard parallel between tiers 100 and 101 north of the base line of the fifth principal meridian.

The standard parallels have been run at varying intervals, the present distance being 24 miles. None at all were used in the earlier surveys. Since public roads are usually built on section and quarter section lines, wherever a north-south road crosses a correction line, there is a "jog" in the road, as a glance at Figure 103 will show.

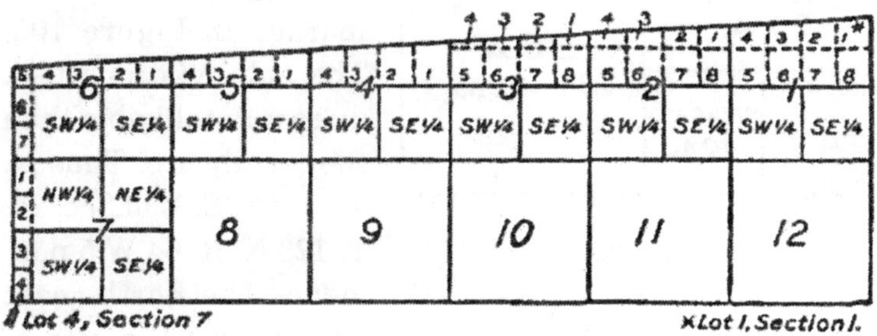

Fig. 104.

Townships Surveyed Northward and Westward. The practice in surveying is to begin at the southeast corner of a township and measure off to the north and west. Thus the sections in the north and west are liable to be larger or smaller than 640 acres, depending upon the accuracy of the survey. In case of a fractional township, made by the intervention of large bodies of water or the meeting of another system of survey or a state line, the sections bear the same numbers they would have if the township were full. Irregular surveys and other causes sometimes make the townships or sections considerably larger than the desired area. In such cases 40 acre lots, or as near that size as possible, appear in the northern row of sections, the other half section remaining as it would otherwise be. These lots may also appear in the western part of a township, and the discrepancy should appear in the western half of each section. This is illustrated in Figure 104.

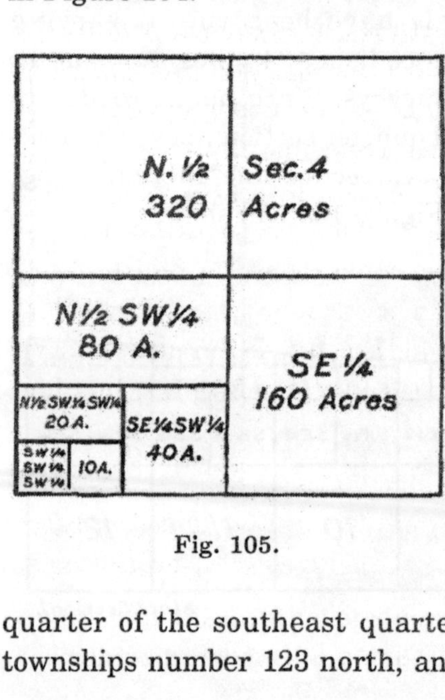

Fig. 105.

Legal Subdivisions of a Section. The legal subdivisions of a section are by halves, quarters, and half quarters. The designation of the portions of a section is marked in Figure 105. The abbreviations look more unintelligible than they really are. Thus N. E. ¼ of S. E. ¼ of Sec. 24, T. 123 N. R. 64 W.5 p.m. means the north- east quarter of the southeast quarter of section 24,in tier of townships number 123 north, and in range 64 west of the fifth principal meridian. Any such description can easily be located on the United States map issued by the General Land Office.

CHAPTER XII

TRIANGULATION IN MEASUREMENT AND SURVEY

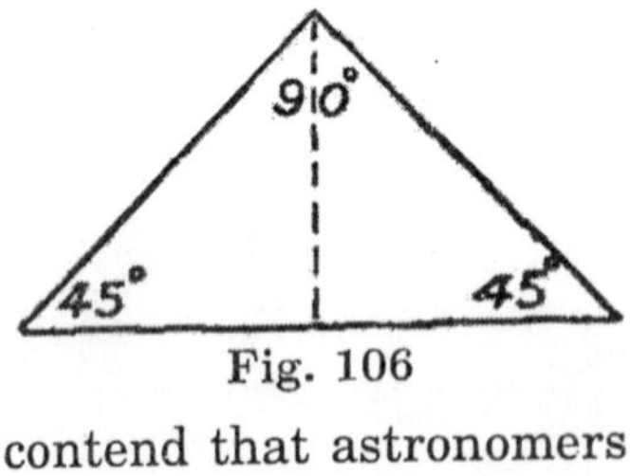

Fig. 106

The ability to measure the distance and size of objects without so much as touching them seems to the child or uneducated person to be a great mystery, if not an impossibility. Uninformed persons sometimes contend that astronomers only guessat the distances and dimensions of the sun, moon, or a planet. The principle of such measurement is very simple and may easily be applied.

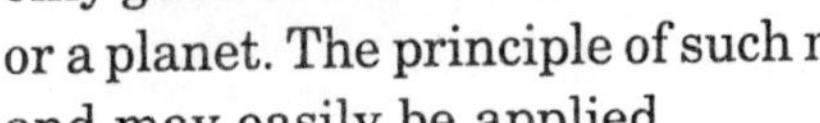

Fig. 107

To Measure the Width of a Stream. Suppose we wish to measure the width of a river, yard, or field without actually crossing it. First make a triangle having two equal sides and one right angle (Fig. 106). Select some easily distinguished point on the farther side, as *X* (Fig.107),and find a convenient point opposite it, as *B*. Now carry the triangle to the right or left of *B* until by sighting you see that the long side is in line with *B* when the short side is in line with *X*. You will then form the triangle *BAX* or *BCX*. It is apparent (by similar triangles) that *AB* or *CB* equals *BX*. Measure off *AB* or *BC* and you will have *BX*, the distance sought. If you measure both to the right and to the left and take the average of the two you will get a more nearly correct result.

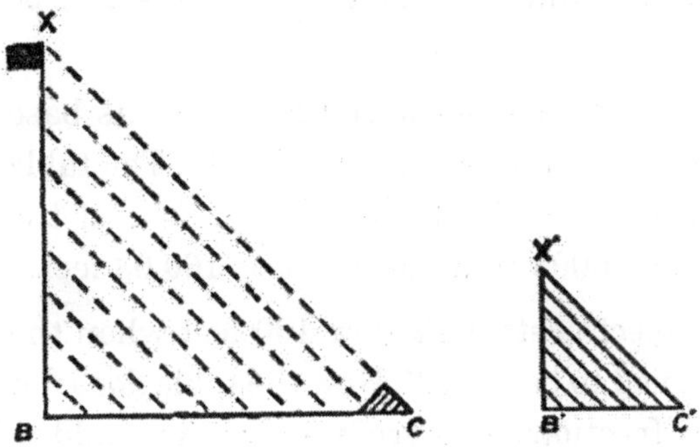

Fig. 108

To Measure the Height of an Object. In a similar manner one may measure the height of a flagstaff or building. Let *X* represent the highest point in the flagstaff (Fig. 108) and place the triangle on or near the ground, with the short side toward *X* and long side level. The distance to the foot of the pole is its height. It is easy to see from this that if we did not have a triangle just as described, say the angle at the point of sighting was less, by measuring that angle and looking up the value of its tangent in a trigonometrical table, one could as easily calculate the height or distance. The angle of the triangle from which sighting was done is 45 °, its tangent is 1.0000, that is, *XB* equals 1.0000 times *BC*. Ifthe angle used were 20 °, instead of 45 °, its tangent would be .3640; that is, *XB* would equal .3640 times *BC*. If the angle were 60°, the tangent would be 1.7321, that is, *XB* would equal that number times *BC*. A complete list of tangents for whole degrees is given in the Appendix. With the graduated quadrant the student can get the noon altitude of the sun (though for this purpose it need not be noon), and by getting the length of shadow and multiplying this by its natural tangent get the height of the object. If it is a building that is thus measured, the distance should be measured from the end of the shadow to the place directly under the point casting the longest shadow measured.

Two examples may suffice to illustrate how this may be done.

1. Say an object casts a shadow 100 feet from its base when the altitude of the sun is observed to be 58°. The table shows the tangent of 58° to be 1.6003. The height of the object, then, must be 1.6003 times 100 feet or 160.03 feet.

2. Suppose an object casts a shadow 100 feet when the sun's height is observed to be 68°12′. Now the table does not give the tangent for fractions of degrees, so we must add to tan 68°1/5 of the difference between the values of tan 68° and tan 69° ($12' = \frac{1}{5}°$)

The table shows that

$\tan 69° = 2.6051$, and

$\tan 68° = 2.4751$, hence the difference = 0.1300.

$\frac{1}{5}$ of .1300 = 0.0260, and since

$\tan 68° = 2.4751$, and we have found that

$\tan 12' = 0.0260$, it follows that

$\tan 68°12' = 2.5011$.

Multiplying 100 feet by this number representing the value of tan 68° 12′

100 feet × 2.5011 = 250.11 feet, answer.

By simple proportion one may also measure the height of an object by the length of the shadow it casts. Let *XB* represent a agstaff and *BC* its shadow on the ground (Fig. 108). Place a ten-foot pole (any other length will do) perpendicularly and measure the length of the shadow it casts and immediately mark the limit of the shadow of the agstaff and measure its length in a level line. Now the length of the flagstaff will bear the same ratio to the length of the pole that the length of the shadow of the flagstaff bears to the length of the shadow of the pole. If the length of the flagstaff's shadow is 60 feet and that of the pole is 6 feet, it is obvious that the former is ten times as high as the latter, or 100 feet high. In formal proportion

BX : B′ X′ : : BC : B′ C′.

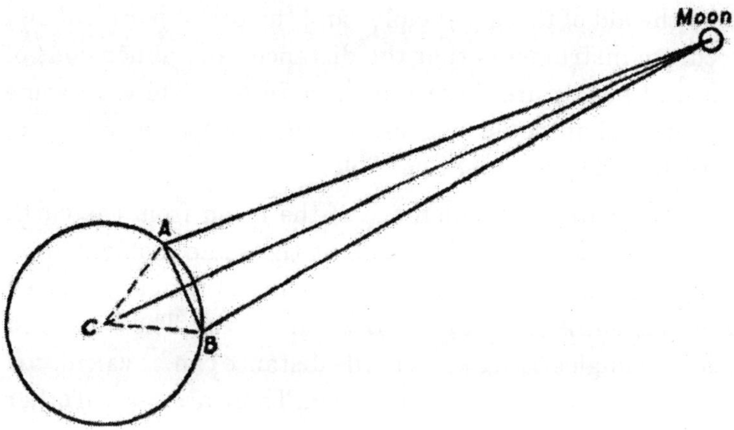

Fig. 109

To Measure the Width of the Moon. To measure the width of the moon if its distance is known. Cut from a piece of paper a circle one inch in diameter and paste it high up on a window in view of the full moon. Find how far the eye must be placed from the disk that the face of the moon may be just covered by the disk. To get this distance it is well to have one person hold the end of a tapeline against the window near the disk and the observer hold the line even with his eye. You then have three elements of the following proportion:

Dist. to disk : dist. to moon : : width of disk : width of moon.

From these elements, multiplying extremes and means and dividing, it is not difficult to get the unknown element, the diameter of the moon. If the student is careful in his measurement and does not forget to reduce all dimensions to the same denomination, either feet or inches, he will be surprised at the accuracy of his measurement, crude though it is.

How Astronomers Measure Sizes and Distances. It is by the aid of these principles and the use of powerful and accurate instruments that the distances and dimensions of celestial bodies are determined, more accurately, in some instances, than would be likely to be done with rod and chain, were such measurement possible.

In measuring the distance of the moon from the earth two observations may be made at the same moment from widely distant points on the earth. Thus a triangle is formed from station A and station B to the moon. The base and included angles being known, the distance can be calculated to the apex of the triangle, the moon. There are several other methods based upon the same general principles, such as two observations from the same point twelve hours apart. Since the calculations are based upon lines conceived to extend to the center of the earth, this is called the geocentric parallax (see Parallax in Glossary). It is impossible to get the geocentric parallax of other stars than the sun because they are so far away that lines sighted to one from opposite sides of the earth are apparently parallel. It is only by making observations six months apart, the diameter of the earth's orbit forming the base of the triangle, that he parallaxes of about forty stars have been determined and even then the departure from the parallel is so exceedingly slight that the distance can be given only approximately. The parallax of stars is called heliocentric, since the base passes through the center of the sun.

SURVEY BY TRIANGULATION

A method very extensively employed for exact measurement of land surfaces is by laying off imaginary triangles across the surface, and by measuring the length of one side and the included angles all other dimensions may be accurately computed. Immense areas in India, Russia, and North America have been thus surveyed. The triangulation surveys of the United States comprise nearly a million square miles extending from the Atlantic to the Pacific. This work has been carried on by the United States Geological Survey for the purpose of mapping the topography and making geological maps, and by the United States Coast and Geodetic Survey.

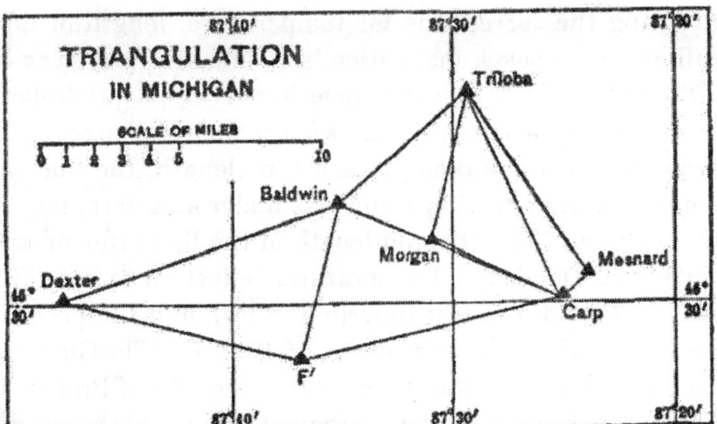

Fig. 110

Determination of Base Line. The surveyor selects two points a few miles apart where the intervening surface is level. The distance between these points is ascertained, great care being used to make it as correct as possible, for this is the base line and all calculations rest for their accuracy upon this distance as it is the only line measured. The following extracts from the Bulletin of the United States Geological Survey on Triangulation, No. 122, illustrate the methods employed. "The Albany base line (in central Texas) is about nine miles in length and was measured twice with a 300-foot steel tape stretched under a tension of 20 pounds. The tape was supported by stakes at intervals of 50 feet, which were aligned and brought to the grade established by more substantial supports, the latter having been previously set in the ground 300 feet apart, and upon which markings of the extremities of the tape were made. The two direct measurements differed by 0.167 foot, but when temperature corrections were applied the resulting discrepancy was somewhat greater, owing possibly to difficulty experienced at the time of measurements in obtaining the true temperature of the tape. The adopted length of the line after applying the corrections for temperature, length of tape, difference on posts, inclination, sag, and sea level, was 45, 793.652 feet." "The base line (near Rapid City, South Dakota) was measured three times with a 300-foot steel tape; temperature was taken at each tape length; the line was supported at each 50 feet and was under a uniform tension of 20 pounds. The adopted length of the line after making corrections for slope, temperature, reduction to sea level, etc., is 25, 796.115 feet (nearly 5 miles), and the probable error of the three measurements is 0.84 inch." "The Gunnison line (Utah) was measured under the direction of Prof. A. H. Thompson, in 1875, the measurement being made by wooden rods carried in a trussed wooden case. These rods were oiled and varnished to prevent absorption of moisture, and their length was carefully determined by comparisons with standard steel rods furnished by the United States Coast and Geodetic Surveys."

Completion of Triangle. From each extremity of the base line a third point is sighted and with an instrument the angle this line forms with the base line is determined. Thus suppose *AB* (Fig. 111) represents the base line. At *A* the angle *CAB* is determined and at *B* the angle *CBA* is determined. Then by trigonometrical tables the lengths of lines *CA* and *BC* are exactly determined. Any one of these lines may now be used as a base for another triangle as with base *AB*. If the first base line is correct, and the angles are determined accurately, and proper allowances are made for elevations and the curvature of the earth, the measurement is very accurate and easily obtained, whatever the intervening obstacles between the points. In some places in the western part of the United States, long lines, sometimes many miles in length, are laid off from one high elevation to another. The longest side thus laid off in the Rocky Mountain region is 183 miles long.

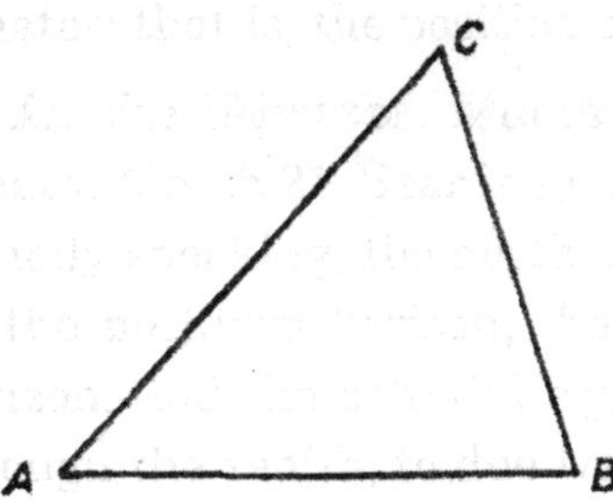

Fig. 111

"On the recent primary triangulation much of the observing has been done at night upon acetylene lamps; directions to the distant light keepers have been sent by the telegraphic alphabet and ashes of light, and the necessary observing towers have been built by a party expert in that kind of work in advance of the observing party."*

Survey of Indian Territory. In March, 1895, Congress provided for the survey of the lands of Indian Territory and

*John F. Hayford, Inspector of Geodetic Work, United States Coast and Geodetic Survey, in a paper relating to Primary Triangulation before the Eighth International Geographic Congress, 1904.

the work was placed in charge of the Director of the Geological Survey instead of being let out on contract as had been previously done. The system of running principal and guide meridians, base and correction parallels, and township and section lines was adopted as usual and since the topographic map was made under the same direction, a survey by triangulation was made at the same time. The generally level character of the country made it possible to make triangles wherever desired, so the "checkerboard" system of townships has superimposed upon it triangles diagonally across the townships. In this way the accurate system of triangulation was used to correct the errors incident to a survey by the chain. Since so many lines were thus laid off and all were made with extreme accuracy, the work of making the contour map was rendered comparatively simple.

CHAPTER XIII

THE EARTH IN SPACE

The Solar System. The group of heavenly bodies to which the earth belongs is called, after its great central sun, the solar system. The members of the solar system are the sun; eight large planets, some having attendant satellites or moons; several hundred smaller planets called asteroids, or planetoids; and occasional comets and meteors. The planets with their satellites, and the asteroids all revolve around the sun in the same direction in elliptical orbits not far from a common plane. Those visible to the naked eye may be seen not far from the ecliptic, the path of the sun in its apparent revolution. The comets and swarms of meteors also revolve around the sun in greatly elongated orbits.

The solar system is *widely separated* from any of the stars, with which the planets should not be confused. If one could fly from the earth to the sun, 93, 000, 000 miles, in a single day, it would take him only a month to reach the orbit of the most distant planet, Neptune, but at that same terrific rate, it would take over seven hundred years to reach the very nearest of the distant stars. If a circle three feet in diameter be made to represent the orbit of the earth, an object over seventy miles away would represent the nearest of the distant stars.

The earth's orbit as seen from the nearest star is as a circle a trie over half an inch in diameter seen at a distance of a mile. Do not imagine that the brightest stars are nearest.

From the foregoing one should not fail to appreciate the immensity of the earth's orbit. It is small only in a relative sense. The earth's orbit is so large that in traveling eighteen and one half miles the earth departs from a perfectly straight line only about *one ninth of an inch*; it is nearly 584, 000, 000 miles in length and the average orbital velocity of the earth is 66, 600 miles per hour.

Sun's Onward Motion. It has been demonstrated that many of the so-called fixed stars are not fixed in relation to each other but have "proper" motions of their own. It is altogether probable that each star has its own motion in the universe. Now the sun is simply one of the stars (see p. 265), and it has been demonstrated that with its system of planets it is moving rapidly, perhaps 40, 000 miles per hour, toward the constellation Hercules. Many speculations are current as to whether our sun is controlled by some other sun somewhat as it controls the planets, and also as to general star systems. Any statement of such conditions with present knowledge is little, if any, more than a guess.

Nebular Hypothesis. Time was when it was considered impious to endeavour to ascertain the processes by which God works "in His mysterious way, His wonders to perform;" and to assign to natural causes and conditions what had been attributed to God's fiat was thought sacrilegious. It is hoped that day has forever passed.

This great theory as to the successive stages and conditions in the development of the solar system, while doubtless faulty in some details, is at present almost the only working hypothesis advanced and "forms the foundation of all the current speculations on the subject." It gives the facts of the solar system a unity and significance scarcely otherwise obtainable.

A theory or a hypothesis, if worthy of serious attention, is always based upon facts. Some of the facts upon which the nebular theory is based are as follows:

1. All of the planets are not far from a common plane.

2. They all revolve around the sun in the same direction.

3. Planetary rotation and revolution are in the same direction, excepting, perhaps, in case of Uranus and Neptune.

4. The satellites revolve around their respective planets in the direction of their rotation and not far from the plane of revolution.

5. All the members seem to be made up of the same kinds of material.

6. Analogy.

a. The nebulæ we see in the heavens have the same general appearances this theory assumes the solar system to have had.

b. The swarms of meteorites making the rings of Saturn are startlingly suggestive of the theory.

c. The gaseous condition of the sun with its corona suggests possible earlier extensions of it. The fact that the sun rotates faster at its equator than at other parts also points toward the nebular theory. The contraction theory of the source of the sun's heat, so generally accepted, is a corollary of the nebular theory.

d. The heated interior of the earth and the characteristics of the geological periods suggest this theory as the explanation.

The Theory. These facts reveal a system intimately related and pointing to a common physical cause. According to the theory, at one time, countless ages ago, all the matter now making up the solar system was in one great cloudlike mass extending beyond the orbit of the most distant planet. This matter was not distributed with uniform density. The greater attraction of the denser portions gave rise to the collection of more matter around them, and just as meteors striking our atmosphere generate by friction the ash of light, sometimes called falling or shooting stars, so the clashing of particles in this nebulous mass generated intense heat.

Rotary Motion. Gradually the whole mass balanced about its center of gravity and a well-defined rotary motion developed. As the great nebulous mass condensed and contracted, it rotated faster and faster. The centrifugal force at the axis of rotation was, of course, zero and increased rapidly toward the equator. The force of gravitation thus being partially counteracted by centrifugal force at the equator, and less and less so at other points toward the axis, the mass attened at the poles. The matter being so extremely thin and tenuous and acted upon by intense heat, also a centrifugal force, it attened out more and more into a disklike form.

As the heat escaped, the mass contracted and rotated faster than ever, the centrifugal force in the outer portion thus increased at a greater rate than did the power of gravitation due to its lessening diameter. Hence, a time came when the centrifugal force of the outer portions exactly balanced the attractive power of gravitation and the rim or outer fragments ceased to contract toward the central mass; and the rest, being nearer the center of gravity, shrank away from these outer portions. The outer ring or ringlike series of fragments, thus left o", continued a rotary motion around the central mass, remaining in essentially the same plane.

Planets Formed from Outlying Portions. Since the matter in the outlying portions, as in the whole mass, was somewhat unevenly distributed, the parts of it consolidated. The greater masses in the outer series hastened by their attraction the lesser particles back of them, retarded those ahead of them, and thus one mass was formed which revolved around the parent mass and rotated on its axis. If this body was not too dense it might collect into the satellites or moons revolving around it. This process continued until nine such rings or lumps had been thrown off, or, rather, *left off*. The many small planets around the sun between the orbit of Mars and that of Jupiter were probably formed from one whose parts were so nearly of the same mass that no one by its preponderating attraction could gather up all into a planet. The explanation of the rings of Saturn is essentially the same.

Conclusion as to the Nebular Hypothesis. This theory, with modifications in detail, forms the basis for much of scientific speculation in subjects having to do with the earth. That it is the ultimate explanation, few will be so hardy as to affirm. Many questions and doubts have been thrown on certain phases recently but it is, in a sense, the point of departure for other theories which may displace it. Perhaps even the best of recent theories to receive the thoughtful attention of the scientific world, the "planetesimal hypothesis," can best be understood in general outline, in terms of the nebular theory.

The Planetesimal Hypothesis. This is a new explanation of the genesis of our solar system which has been worked out by Professors Chamberlin and Moulton of the University of Chicago, and is based upon a very careful study of astronomical facts in the light of mathematics and astrophysics. It assumes the systems to have been evolved from a spiral nebula, similar to the most common form of nebulæ observed in the heavens. It is supposed that the nebulous condition may have been caused by our sun passing so near a star that the tremendous tidal strain caused the eruptive prominences (which the sun shoots out at frequent intervals) to be much larger and more vigorous than usual, and that these, when projected far out, were pulled forward by the passing star and given a revolutionary course about the sun. The arms of spiral nebulæ have knots of denser matter at intervals which are supposed to be due to special explosive impulses and to become the centers of accretion later. The material thus shot out was very hot at first, but soon cooled into discrete bodies or particles which moved independently about the sun like planets (hence the term *planetesimal*). When their orbits crossed or approached each other, the smaller particles were gathered into the knots, and these ultimately grew into planets. Less than one seven-hundredth of the sun was necessary to form the planets and satellites.

This hypothesis differs from the nebular hypothesis in a number of important particulars. The latter assumes the earth to have been originally in a highly heated condition, while under the planetesimal hypothesis the earth may have been measurably cool at the surface at all times, the interior heat being due to the compression caused by gravity. The nebular hypothesis views the atmosphere as the thin remnant of voluminous original gases, whereas the new hypothesis conceives the atmosphere to have been gathered gradually about as fast as consumed, and to have come in part from the heated interior, chiefly by volcanic action, and in part from outer space. The oceans, according to the old theory, were condensed from the great masses of original aqueous vapors surrounding the earth; according to the new theory the water was derived from the same sources as the atmosphere. According to the planetesimal hypothesis the earth, as a whole, has been solid throughout its history, and never in the molten state assumed in the nebular hypothesis.

Solar System not Eternal. Of one thing we may be reasonably certain, the solar system is not an eternal one. When we endeavor to extend our thought and imagination backward toward "the beginning," it is only toward creation; when forward, it is only *toward* eternity.

> "Thy kingdom is an everlasting kingdom,
> And thy dominion endureth throughout all generations."
>
> —Psalms, 145, 13.

MATHEMATICAL GEOGRAPHY OF THE PLANETS, MOON, AND SUN

The following brief sketches of the mathematical geography of the planets give their conditions in terms corresponding to those applied to the earth. The data and comparisons with the earth are only approximate. The more exact figures are found in the table at the end of the chapter.

Striving for vividness of description occasionally results in language which implies the possibility of human inhabitancy on other celestial bodies than the earth, or suggests interplanetary locomotion (see p. 304). Such conditions *exist only in the imagination.* An attempt to exclude astronomical facts not bearing upon the topic in hand and not consistent with the purpose of the study, makes necessary the omission of some of the most interesting facts. For such information the student should consult an astronomy. The beginner should learn the names of the planets in the order of their nearness to the sun. Three minutes repetition, with an occasional review, will fix the order:

Mercury, Venus, Earth, Mars, Asteroids,
Jupiter, Saturn, Uranus, Neptune.

There are obvious advantages in the following discussion in not observing this sequence, taking Mars first, then Venus, etc.

MARS

Form and Dimensions. In form Mars is very similar to the earth, being slightly more attened toward the poles. Its mean diameter is 4, 200 miles, a little more than half the earth's. A degree of latitude near the equator is 36.6 miles long, getting somewhat longer toward the poles as in case of terrestrial latitudes.

Mars has a little less than one third the surface of the earth, has one seventh the volume, weighs but one ninth as much, is three fourths as dense, and an object on its surface weighs about two fifths as much as it would here. A man weighing one hundred and fifty pounds on the earth would weigh only fifty-seven pounds on Mars, could jump two and one half times as high or far, and could throw a stone two and one half times the distance he could here.* A pendulum clock taken from the earth to Mars would lose nearly nine hours in a day as the pendulum would tick only about seven elevenths as fast there. A watch, however, would run essentially the same there as here. As we shall see presently, either instrument would have to be adjusted in order to keep Martian time as the day there is longer than ours.

Rotation. Because of its well-marked surface it has been possible to ascertain the period of rotation of Mars with very great precision. Its sidereal day is 24 h. 37 m. 22.7 s. The solar day is 39 minutes longer than our solar day and owing to the greater ellipticity of its orbit the solar days vary more in length than do ours.

Revolution and Seasons. A year on Mars has 668 Martian days,† and is nearly twice as long as ours. The orbit is much more elliptical than that of the earth, perihelion being 26, 000, 000 miles nearer the sun than aphelion. For this reason there is a marked change in the amount of heat received when Mars is at those two points, being almost one and one half times as much when in perihelion as when in aphelion. The northern summers occur when Mars is in aphelion, so that hemisphere has longer, cooler summers

*He could not throw the stone any swifter on Mars than he could on the earth; gravity there being so much weaker, the stone would move farther before falling to the surface.

† *Mars*, by Percival Lowell.

and shorter and warmer winters than the southern hemisphere.

NORTHERN HEMISPHERE		SOUTHERN HEMISPHERE	
Spring	191 days	Spring	149 days
Summer	181 days	Summer	147 days
Autumn	149 days	Autumn	191 days
Winter	147 days	Winter	181 days

Zones. The equator makes an angle of 24°50′ with the planet's ecliptic (insteade of 23°27′ as with us) so the change in seasons and zones is very similar to ours, the climate, of course, being vastly different, probably *very cold* because of the rarity of the atmosphere (about the same as on our highest mountains) and absence of oceans. The distance from the sun, too, makes a great difference in climate. Being about one and one half times as far from the earth, the sun has an apparent diameter only two thirds as great and only four ninths as much heat is received over a similar area.

Satellites. Mars has two satellites or moons. Since Mars was the god of war of the Greeks these two satellites have been given the Greek names of Deimos and Phobos, meaning "dread" and "terror," appropriate for "dogs of war." They are very small, only six or seven miles in diameter. Phobos is so near to Mars (3,750 miles from the surface) that it looks almost as large to a Martian as our moon does to us, although not nearly so bright. Phobos, being so near to Mars, has a very swift motion around the planet, making more than three revolutions around it during a single Martian day. Now our moon travels around the earth from west to east but only about 13° in a day, so because of the earth's rotation the moon rises in the east and sets in the west. In case of Phobos, it revolves faster than the planet rotates and thus rises in the west and sets in the east. Thus if Phobos rose in the west at sunset in less than three hours it would be at meridian height and show first quarter, in five and one half hours it would set in the east somewhat past the full, and before sunrise would rise again in the west almost at the full again. Deimos has a sidereal period of 30.3 hours and thus rises in the east and sets in the west, the period from rising to setting being 61 hours.

VENUS

Form and Dimensions. Venus is very nearly spherical and has a diameter of 7, 700 miles, very nearly that of the earth, so its latitude and longitude are very similar to ours. Its surface gravity is about $\frac{9}{10}$ that of the earth. A man weighing 150 pounds here would weigh 135 pounds there.

Revolution. Venus revolves around the sun in a period of 225 of our days, probably rotating once on the journey, thus keeping essentially the same face toward the sun. The day, therefore, is practically the same as the year, and the zones are two, one of perpetual sunshine and heat and the other of perpetual darkness and cold. Its atmosphere is of nearly the same density as that of the earth. Being a little more than seven tenths the distance of the earth from the sun, that blazing orb seems to have a diameter nearly one and one half times as great and pours nearly twice as much light and heat over a similar area. Its orbit is more nearly circular than that of any other planet.

JUPITER

Form and Dimensions. After Venus, this is the brightest of the heavenly bodies, being immensely large and having very high reecting power. Jupiter is decidedly oblate.

Its equatorial diameter is 90,000 miles and its polar diameter is 84,200 miles. Degrees of latitude near the equator are thus nearly 785 miles long, increasing to over 800 miles near the pole. The area of the surface is 122 times that of the earth, its volume 1,355, its mass or weight 317, and its density about one fourth.

Surface Gravity. The weight of an object on the surface of Jupiter is about two and two thirds times its weight here. A man weighing 150 pounds here would weigh 400 pounds there but would find he weighed nearly 80 pounds more near the pole than at the equator, gravity being so much more powerful there. A pendulum clock taken from the earth to Jupiter would gain over nine hours in a day and would gain or lose appreciably in changing a single degree of latitude because of the oblateness of the planet.

Rotation. The rotation of this planet is very rapid, occupying a little less than ten hours, and some portions seem to rotate faster than others. It seems to be in a molten or liquid state with an extensive envelope of gases, eddies and currents of which move with terrific speed. The day there is very short as compared with ours and a difference of one hour in time makes a difference of over 36° in longitude, instead of 15. as with us. Their year being about 10, 484 of their days, their solar day is only a few seconds longer than their sidereal day.

Revolution. The orbit of Jupiter is elliptical, perihelion being about 42, 000, 000 miles nearer the sun than aphelion. Its mean distance from the sun is 483, 000, 000 miles, abouts five times that of the earth. The angle its equator forms with it ecliptic is only 3° so there is little change in seasons. The vertical ray of the sun never gets more than 3° from the equator, and the torrid zone is 6° wide. The circle of illumination is never more than 3° from or beyond a pole so the frigid zone is only 3° wide. The temperate* zone is 84° wide.

Jupiter has seven moons.

SATURN

Form and Dimensions. The oblateness of this planet is even greater than that of Jupiter, being the greatest of the planets. Its mean diameter is about 73,000 miles. It, therefore, has 768 times the volume of the earth and 84 times the surface. Its density is the lowest of the planets, only about one eighth as dense as the earth. Its surface gravity is only slightly more than that of the earth, varying, however, 25 per cent from pole to equator.

Rotation. Its sidereal period of rotation is about 10 h. 14 m., varying slightly for different portions as in case of Jupiter. The solar day is only a few seconds longer than the sidereal day.

Revolution. Its average distance from the sun is 866, 000,000 miles, varying considerably because of its ellipticity. It revolves about the sun in 29.46 of our years, thus the annual calendar must comprise 322,777 of the planet's days.

The inclination of Saturn's axis makes an angle of 27° between the planes of its equator and its ecliptic. Thus the vertical ray sweeps over 54° giving that width to its torrid zone, 20° to the frigid, and 36° to the temperate. Its ecliptic and our ecliptic form an angle of 2.5° so we always see the planet very near the sun's apparent path.

*These terms are purely relative, meaning, simply, the zone on Jupiter corresponding in position to the temperate zone on the earth. The inappropriateness of the term may be seen in the fact that Jupiter is intensely heated, so that its surface beneath the massive hot vapors surrounding it is probably molten.

Saturn has surrounding its equator immense disks, of thin, gauzelike rings, extending out nearly 50, 000 miles from the surface. These are swarms of meteors or tiny moons, swinging around the planet in very nearly the same plane, the inner ones moving faster than the outer ones and being so very minute that they exert no appreciable attractive inuence upon the planet.

In addition to the rings, Saturn has ten moons.

URANUS

Form and Dimensions. This planet, which is barely visible to the unaided eye, is also decidedly oblate, nearly as much so as Saturn. Its mean diameter is given as from 34, 900 miles to 28,500 miles. Its volume, on basis of the latter (and latest) figures, is 47 times that of the earth. Its density is very low, about three tenths that of the earth, and its surface gravity is about the same as ours at the equator, increasing somewhat toward the pole.

Nothing certain is known concerning its rotation as it has no distinct markings upon its surface. Consequently we know nothing as to the axis, equator, days, calendar, or seasons.

Its mean distance from the sun is 19.2 times that of the earth and its sidereal year 84.02 of our years.

Uranus has four satellites swinging around the planet in very nearly the same plane at an angle of 82.2° to the plane of the orbit. They move from west to east around the planet, not for the same reason Phobos does about Mars, but probably because the axis of the planet, the plane of its equator, and the plane of these moons has been tipped 97.8° from the plane of the orbit and the north pole has been tipped down below or south of the ecliptic, becoming the south pole, and giving a backward rotation to the planet and to its moons.

NEPTUNE

Neptune is the most distant planet from the sun, is probably somewhat larger than Uranus, and has about the same density and slightly greater surface gravity. Owing to the absence of definite markings nothing is known as to its rotation. Its one moon, like those of Uranus, moves about the planet from west to east in a plane at an angle of 34° 48′ to its ecliptic, and its backward motion suggests a similar explanation, the inclination of its axis is more than 90° from the plane of its ecliptic.

MERCURY

This is the nearest of the planets to the sun, and as it never gets away from the sun more than about the width of forty suns (as seen from the earth), it is rarely visible and then only after sunset in March and April or before sunrise in September and October.

Form and Dimensions. Mercury has about three eighths the diameter of the earth, one seventh of the surface, and one eighteenth of the volume. It probably has one twentieth of the mass, nine tenths of the density, and a little less than one third of the surface gravity.

Rotation and Revolution. It is believed that Mercury rotates once on its axis during one revolution. Owing to its elliptical orbit it moves much more rapidly when near perihelion than when near aphelion, and thus the sun loses as compared with the average position, just as it does in the case of the earth, and sweeps eastward about 23½° from its average position. When in aphelion it gains and sweeps westward a similar amount. This shifting eastward making the sun "slow" and westward making the sun "fast" is called libration.

Thus there are four zones on Mercury, vastly different from ours, indeed, they are not zones (belts) in a terrestrial sense.

a. An elliptical central zone of perpetual sunshine, extending from pole to pole and 133° in longitude. In this zone the vertical ray shifts eastward 23½° and back again in the short summer of about 30 days, and westward a similar extent during the longer winter of about 58 days. Two and one half times as much heat is received in the summer, when in perihelion, as is received in the winter, when in aphelion. Thus the eastward half of this zone has hotter summers and cooler winters than does the western half. Places along the eastern and western margin of this zone of perpetual sunshine see the sun on the horizon in winter and only 23½° high in the summer.

b. An elliptical zone of perpetual darkness, extending from pole to pole and 133° wide from east to west.

c. Two elliptical zones of alternating sunshine and darkness (there being practically no atmosphere on Mercury, there is no twilight there), each extending from pole to pole and 47° wide. The eastern of these zones has hotter summers and cooler winters than the western one has.

THE MOON

Form and Dimensions. The moon is very nearly spherical and has a diameter of 2,163 miles, a little over one fourth that of the earth, its volume one forty-fourth, its density three fifths, its mass $\frac{10}{815}$, and its surface gravity one sixth that upon the earth. A pendulum clock taken there from the earth would tick so slowly that it would require about sixty hours to register one of our days. A degree of latitude (or longitude at its equator) is a little less than nineteen miles long.

Rotation. The moon rotates exactly once in one revolution around the earth, that is, keeps the same face toward the earth, but turns different sides toward the sun once each month.

Thus what we call a sidereal month is for the moon itself a sidereal day, and a synodic month is its solar day. The latter is 29.5306 of our days, which makes the moon's solar day have 708 h. 44 m. 3.8 s. If its day were divided into twenty-four parts as is ours, each one would be longer than a whole day with us.

Revolution and Seasons. The moon's orbit around the sun has essentially the same characteristics as to perihelion, aphelion, longer and shorter days, etc., as that of the earth. The fact that the moon goes around the earth does not materially affect it from the sun's view point. To illustrate the moon's orbit about the sun, draw a circle 78 inches in diameter. Make 26 equidistant dots in this circle to represent the earth for each new and full moon of the year. Now for each new moon make a dot one twentieth of an inch toward the center (sun) from every other dot representing the earth, and for every full moon make a dot one twentieth of an inch beyond the alternate ones. These dots representing the moon, if connected, being never more than about one twentieth of an inch from the circle, will not vary materially from the circle representing the orbit of the earth, and the moon's orbit around the sun will be seen to have in every part a concave side toward the sun.

The solar day of the moon being 29.53 of our days, its tropical year must contain as many of those days as that number is contained times in 365.25 days or about 12.4 days. The calendar for the moon does not have anything corresponding to our month, unless each day be treated as a month, but has a year of 12.4 long days of nearly 709 hours each. The exact length of the moon's solar year being 12.3689 d., its calendar would have the peculiarity of having one leap year in every three, that is, two years of 12 days each and then one of 13 days, with an extra leap year every 28 years.

The earth as seen from the moon is much like the moon as seen from the earth, though very much larger, about four times as broad. Because the moon keeps the same face constantly toward the earth, the latter is visible to only a little over half of the moon. On this earthward side our planet would be always visible, passing through precisely the same phases as the moon does for us, though in the opposite order, the time of our new moon being "full earth" for the moon. So brightly does our earth then illuminate the moon that when only the faint crescent of the sunshine is visible to us on the rim of the moon, we can plainly see the "earth shine" on the rest of the moon's surface which is toward us.

Zones. The inclination of the plane of the moon's equator to the plane of the ecliptic is 1° 32′ (instead of 23° 27′ as in the case of the earth). Thus its zone corresponding to our torrid* zone is 3° 4′ wide, the frigid zone 1° 32′, and the temperate zones 86° 560′.

Absence of Atmosphere. The absence of an atmosphere on the moon makes conditions there vastly different from those to which we are accustomed. Sunrise and sunset show no crimson tints nor beautiful coloring and there is no twilight. Owing to the very slow rotation of the moon, 709 hours from sun-noon to sun-noon, it takes nearly an hour

*Again we remind the reader that these terms are not appropriate in case of other celestial bodies than the earth. The moon has almost no atmosphere to retain the sun's heat during its long night of nearly 354 hours and its dark surface must get exceedingly cold, probably several hundred degrees below zero.

for the disk of the sun to get entirely above the horizon on the equator, from the time the first glint of light appears, and the time of sunset is equally prolonged; as on the earth, the time occupied in rising or setting is longer toward the poles of the moon. The stars do not twinkle, but shine with a clear, penetrating light. They may be seen as easily in the daytime as at night, even those very near the sun. Mercury is thus visible the most of the time during the long daytime of 354 hours, and Venus as well. Out of the direct rays of the sun, pitch darkness prevails. Thus craters of the volcanoes are very dark and also cold. In the tropical portion the temperature probably varies from two or three hundred degrees below zero at night to exceedingly high temperatures in the middle of the day. During what is to the moon an eclipse of the sun, which occurs whenever we see the moon eclipsed, the sun's light shining through our atmosphere makes the most beautiful of coloring as viewed from the moon. The moon's atmosphere is so rare that it is incapable of transmitting sound, so that a deathlike silence prevails there. Oral conversation is utterly impossible and the telephone and telegraph as we have them would be of no use whatever. Not a drop of water exists on that cold and cheerless satellite.

Perhaps it is worth noting, in conclusion, that it is believed that our own atmosphere is but the thin remnant of dense gases, and that in ages to come it will get more and more rarified, until at length the earth will have the same conditions as to temperature, silence, etc., which now prevail on the moon.

THE SUN

Dimensions. The diameter is 866, 500 miles, nearly four times the distance of the moon from the earth. Its surface area is about 12, 000 times that of the earth, and its volume over a million times. Its density is about one fourth that of the earth, its mass 332, 000 times, and its surface gravity is 27.6 times our earth's. A man weighing 150 pounds here would weigh over two tons there, his arm would be so heavy he could not raise it and his bony framework could not possibly support his body. A pendulum clock there would gain over a hundred hours in a day, so fast would the attraction of the sun draw the pendulum.

Rotation. The sun rotates on its axis in about $25\frac{1}{3}$ of our days, showing the same portion to the earth every $27\frac{1}{4}$ days.This rate varies for different portions of the sun, its equator rotating considerably faster than higher latitudes. The direction of its rotation is from west to east from the sun's point of view, though as viewed from the earth the direction is from our east to our west. The plane of the equator forms an angle of about 26° with the plane of our equator, though only about $7\frac{1}{4}°$ with the plane of the ecliptic.

When we realize that the earth, as viewed from the sun, is so tiny that it receives not more than one billionth of its light and heat, we may form some idea of the immense flood of energy it constantly pours forth.

The Sun a Star. "The word 'star' should be omitted from astronomical literature. It has no astronomic meaning. Every star visible in the most penetrating telescope is a hot sun. They are at all degrees of heat, from dull red to the most terrific white heat to which matter can be subjected. Leaves in a forest, from swelling bud to the `sere and yellow,' do not present more stages of evolution. A few suns that have been weighed, contain less matter than our own; some of equal mass; others are from ten to twenty and thirty times more massive, while a few are so immensely more massive that all hopes of comparison fail.

"Every sun is in motion at great speed, due to the attraction and counter attraction of all the others. They go in every direction. Imagine the space occupied by a swarm of bees to be magnified so that the distance between each bee and its neighbor should equal one hundred miles. The insects would fly in every possible direction of their own volition. Suns move in every conceivable direction, not as they will, but in abject servitude to gravitation. They must obey the omnipresent force, and do so with mathematical accuracy." From "New Conceptions in Astronomy," by Edgar L. Larkin, in *Scientific American*, February 3, 1906.

SOLAR SYSTEM TABLE

Object	Symbol	Mean Diameter (miles)	Sidereal Day	As compared with the earth*					
				Heat per Unit Area	Density	Mass	Surface Gravity	Sidercal Year	Dist. from ⊙
Mercury	☿	3,000	88 days	6.800	0.85	0.048	0.330	0.24	0.4
Venus	♀	7,700	225 days	1.900	0.94	0.820	0.900	0.62	0.7
Earth	♁	7,918	*	1.000	1.00	1.000	1.000	1.00	1.0
Mars	♂	4,230	24h 37m 22.7s	0.440	0.73	0.110	0.380	1.88	1.5
Jupiter	♃	88,000	9h 55m	0.040	0.23	317.000	2.650	11.86	5.2
Saturn	♄	73,000	10h 14m	0.010	0.13	95.000	1.180	29.46	9.5
Uranus	♅	31,700	?	0.003	0.31	14.600	1.110	84.02	19.2
Neptune	♆	32,000	?	0.001	0.34	17.000	1.250	164.78	30.1
Sun	⊙	866,400	25d 7h 48m		0.25	332,000.000	27.650		
Moon	☾	2,163	27d 7h 43m		0.61	0.012	0.166		

*The dimensions of the earth and other data are given in the table of geographical constants p. 308.

CHAPTER-XIV

HISTORICAL SKETCH

THE FORM OF THE EARTH

While various views have been held regarding the form of the earth, those worthy of attention* may be grouped under four general divisions.

I. The Earth Flat. Doubtless the universal belief of primitive man was that, save for the irregularities of mountain, hill, and valley, the surface of the earth is at. In all the earliest literature that condition seems to be assumed. The ancient navigators could hardly have failed to observe the apparent convex surface of the sea and very ancient literature as that of Homer alludes to the bended sea. This, however, does not necessarily indicate a belief in the spherical form of the earth.

Although previous to his time the doctrine of the spherical form of the earth had been advanced, Herodotus (born about 484 b.c., died about 425 b.c.) did not believe in it and scouted whatever evidence was advanced in its favor. Thus in giving the history of the Ptolemys, kings of Egypt,

*As for modern, not to say recent, pseudo-scientists and alleged divine revealers who contend for earths of divers forms, the reader is referred to the entertaining chapter entitled "Some Cranks and their Crotchets" in John Fiske's *A Century of Science,* also the footnote on pp. 267–268, Vol. I, of his *Discovery of America.*

he relates the incident of Ptolemy Necho (about 610–595 b.c.) sending Ph nician sailors on a voyage around Africa, and after giving the sailors' report that they saw the *sun to the northward* of them, he says, "I, for my part, do not believe them." Now seeing the sun to the northward is the most logical result if the earth be a sphere and the sailors went south of the equator or south of the tropic of Cancer in the northern summer.

Ancient travelers often remarked the apparent sinking of southern stars and rising of northern stars as they traveled northward, and the opposite shifting of the heavens as they traveled southward again. In traveling eastward or westward there was no displacement of the heavens and travel was so slow that the difference in time of sunrise or star-rise could not be observed. To infer that the earth is curved, at least in a north-south direction, was most simple and logical. It is not strange that some began to teach that the earth is a cylinder. Anaximander (about 611–547 b.c.), indeed, did teach that it is a cylinder* and thus prepared the way for the more nearly correct theory.

II. The Earth a Sphere. The fact that the Chaldeans had determined the length of the tropical year within less than a minute of its actual value, had discovered the precession of the equinoxes, and could predict eclipses over two thousand years before the Christian era and that in China similar facts were known, possibly at an earlier period, would indicate that doubtless many of the astronomers of those very ancient times had correct theories as to the form and motions of the earth. So far as history has left any positive record, however, Pythagoras (about 582–507 b.c.),

*According to some authorities he taught that the earth is a sphere and made terrestrial and celestial globes. See Ball's History of Mathematics, p. 18.

a Greek* philosopher, seems to have been the first to advance the idea that the earth is a sphere. His theory being based largely upon philosophy, nothing but a perfect sphere would have answered for his conception. He was also the first to teach that the earth rotates† on its axis and revolves about the sun.

Before the time of Pythagoras, Thales (about 640–546 b.c.), and other Greek philosophers had divided the earth into five zones, the torrid zone being usually considered so fiery hot that it could not be crossed, much less inhabited. Thales is quoted by Plutarch as believing that the earth is a sphere, but it seems to have been proved that Plutarch was in error. Many of the ancient philosophers did not dare to teach publicly doctrines not commonly accepted, for fear of punishment for impiety. It is possible that his private teaching was different from his public utterances, and that after all Plutarch was right.

Heraclitus, Plato, Eudoxus, Aristotle and many others in the next two centuries taught the spherical form of the earth, and, perhaps, some of them its rotation. Most of them, however, thought it not in harmony with a perfect universe, or that it was impious, to consider the sun as predominant and so tught the geocentric theory.

The first really scientific attempt to calculate the size of the earth was by Eratosthenes (about 275–195 b.c.). He was the keeper of the royal library at Alexandria, and made many astronomical measurements and calculations of very great

*Sometimes called a Phcenician.

†Strictly speaking, Pythagoras seems to have taught that both sun and earth revolved about a central fire and an opposite earth revolved about the earth as a shield from the central fire. This rather complicated machinery offered so many difficulties that his followers abandoned the idea of the central fire and "opposite earth" and had the earth rotate on its own axis.

value, not only for his own day but for ours as well. Syene, the most southerly city of the Egypt of his day, was situated where the sundial cast no shadow at the summer solstice. Measuring carefully at Alexandria, he found the noon sun to be one fiftieth of the circumference to the south of overhead. He then multiplied the distance between Syene and Alexandria, 5, 000 stadia by 50 and got the whole circumference of the earth to be 250, 000 stadia. The distance between the cities was not known very accurately and his calculation probably contained a large margin of error, but the exact length of the Greek stadium of his day is not known* and we cannot tell how near the truth he came.

Any sketch of ancient geography would be incomplete without mention of Strabo (about 54 b.c.–21 a.d.) who is sometimes called the "father of geography." He believed the earth to be a sphere at the center of the universe. He continued the idea of the five zones, used such circles as had commonly been employed by astronomers and geographers before him, such as the equator, tropics, and polar circles. His work was a standard authority for many centuries.

About a century after the time of Eratosthenes, Posidonius, a contemporary of Strabo, made another measurement, basing his calculations upon observations of a star instead of the sun, and getting a smaller circumference, though that of Eratosthenes was probably too small. Strabo, Hipparchus, Ptolemy and many others made estimates as to the size of the earth, but we have no record of any further measurements with a view to exact calculation until about 814 a.d. when the Arabian caliph Al-Mamoum sent astronomers and surveyors northward and southward, carefully measuring the distance until each party found a star to have shifted to the south or north one degree. This

*The most reliable data seem to indicate the length of the stadium was 606¾ feet.

distance of two degrees was then multiplied by 180 and the whole circumference obtained.

The period of the dark ages was marked by a decline in learning and to some extent a reversion to primitive conceptions concerning the size, form, or mathematical properties of the earth. Almost no additional knowledge was acquired until early in the seventeenth century. Perhaps this statement may appear strange to some readers, for this was long after the discovery of America by Columbus. It should be borne in mind that his voyage and the resulting discoveries and explorations contributed nothing directly to the knowledge of the form or size of the earth. That the earth is a sphere was generally believed by practically all educated people for centuries before the days of Columbus. The Greek astronomer Cleomedes, writing over a thousand years before Columbus was born, said that all competent persons excepting the Epicureans accepted the doctrine of the spherical form of the earth.

In 1615 Willebrord Snell, professor of mathematics at the University of Leyden, made a careful triangular survey of the level surfaces about Leyden and calculated the length of a degree of latitude to be 66.73 miles. A recalculation of his data with corrections which he suggested gives the much more accurate measurement of 69.07 miles. About twenty years later, an Englishman named Richard Norwood made measurements and calculations in southern England and gave 69.5 as the length of a degree of latitude, the most accurate measurement up to that time.

It was about 1660 when Isaac Newton (1642–1727) discovered the laws of gravitation, but when he applied the laws to the motions of the moon his calculations did not harmonize with what he assumed to be the size of the earth. About 1671 the French astronomer, Jean Picard, by the use of the telescope, made very careful measurements of a little over a degree of longitude and obtained a close approximation to its length. Newton, learning of the measurement of Picard, recalculated the mass of the earth and motions of the moon and found his law of gravitation as the satisfactory explanation of all the conditions. Then, in 1682, after having patiently waited over twenty years for this confirmation, he announced the laws of gravitation, one of the greatest discoveries in the history of mankind. We find in this an excellent instance of the interdependence of the sciences. The careful measurement of the size of the earth has contributed immensely to the sciences of astronomy and physics.

III. The Earth an Oblate Spheroid. From the many calculations which Newton's fertile brain could now make, he soon was enabled to announce that the earth must be, not a true sphere, but an oblate spheroid. Christian Huygens, a celebrated contemporary of Newton, also contended for the oblate form of the earth, although not on the same grounds as those advanced by Newton.

In about 1672 the trip of the astronomer Richer to French Guiana, South America, and his discovery that pendulums swing more slowly there (see the discussion under the topic The Earth an Oblate Spheroid, p. 27), and the resulting conclusion that the earth is not a true sphere, but is attened toward the poles, gave a new impetus to the study of the size of the earth and other mathematical properties of it.

Over half a century had to pass, however, before the true significance of Richer's discovery was apparent to all or generally accepted. An instance of a commonly accepted reason assigned for the shorter equatorial pendulum is the following explanation which was given to James II of England when he made a visit to the Paris Observatory in 1697. "While Jupiter at times appears to be not perfectly spherical, we may bear in mind the fact that the theory of the earth being attened is sufficiently disproven by the circular shadow which the earth throws on the moon. The apparent necessary shortening of the pendulum toward the south is really only a correction for the expansion of the pendulum in consequence of the higher temperature." It is interesting to note that if this explanation were the true one, the average temperature at Cayenne would have to be 43° above the boiling point.

Early in the eighteenth century Giovanni Cassini, the astronomer in charge of the Paris Observatory, assisted by his son, continued the measurement begun by Picard and came to the conclusion that the earth is a prolate spheroid. A warm discussion arose and the Paris Academy of Sciences decided to settle the matter by careful measurements in polar and equatorial regions.

In 1735 two expeditions were sent out, one into Lapland and the other into Peru. Their measurements, while not without appreciable errors, showed the decided difference of over half a mile for one degree and demonstrated conclusively the oblateness of a meridian and, as Voltaire wittily remarked at the time, "flattened the poles and the Cassinis."

The calculation of the oblateness of the earth has occupied the attention of many since the time of Newton. His calculation was $\frac{1}{230}$; that is, the polar diameter was $\frac{1}{230}$ shorter than the equatorial. Huygens estimated the attening to be about $\frac{1}{500}$. The most commonly accepted spheroid representing the earth is the one calculated in 1866 by A. R. Clarke, for a long time at the head of the English Ordnance Survey (see p. 29). Purely astronomical calculations, based upon the effect of the bulging of the equator upon the motion of the moon, seem to indicate slightly less oblateness than that of General Clarke. Professor William Harkness, formerly astronomical director of the United States Naval Observatory, calculated it to be very nearly $\frac{1}{230}$.

IV. The Earth a Geoid. During recent years many careful measurements have been made on various portions of the globe and extensive pendulum tests given to ascertain the force of gravity. These measurements demonstrate that the earth is not a true sphere; is not an oblate spheroid; indeed, its figure does not correspond to that of any regular or symmetrical geometric form. As explained in Chapter II, the equator, parallels, and meridians are not true circles, but are more or less elliptical and wavy in outline. The extensive triangulation surveys and the application of astrophysics to astronomy and geodesy make possible, and at the same time make imperative, a careful determination of the exact form of the geoid.

THE MOTIONS OF THE EARTH

The Pythagoreans maintained as a principle in their philosophy that the earth rotates on its axis and revolves about the sun. Basing their theory upon *a priori* reasoning, they had little better grounds for their belief than those who thought otherwise. Aristarchus (about 310–250 b.c.), a Greek astronomer, seems to have been the first to advance the heliocentric theory in a systematic manner and one based upon careful observations and calculations. From this time, however, until the time of Copernicus, the geocentric theory was almost universally adopted.

The geocentric theory is often called the Ptolemaic system from Claudius Ptolemy (not to be confused with ancient Egyptian kings of the same name), an Alexandrian astronomer and mathematician, who seems to have done most of his work about the middle of the second century, a.d. He seems to have adopted, in general, the valuable astronomical calculations of Hipparchus (about 180–110 b.c.). The system is called after him because he compiled so much of the observations of other astronomers who had preceded him and invented a most ingenious system of "cycles," "epicycles," "deferents," "centrics," and "eccentrics" (now happily swept away by the Copernican system) by which practically all of the known facts of the celestial bodies and their movements could be accounted for and yet assume the earth to be at the center of the universe.

Among Ptolemy's contributions to mathematical geography were his employment of the latitude and longitude of places to represent their positions on the globe (a scheme probably invented by Hipparchus), and he was the first to use the terms "meridians of longitude" and "parallels of latitude." It is from the Latin translation of his subdivisions of degrees that we get the terms "minutes" and "seconds" (for centuries the division had been followed, originating with the Chaldeans. See p.142). The sixty subdivisions he called first small parts; in Latin, *"minutæ primæ,"* whence our term "minute." The sixty subdivisions of the minute he called second small parts; in Latin, *"minutæ secundæ,"* whence our term "second."

The Copernican theory of the solar system, which has universally displaced all others, gets its name from the Polish astronomer Nicolas Copernicus (1473–1543). He revived the theory of Aristarchus, and contended that the earth is not at the center of the solar system, but that the sun is, and planets all revolve around the sun. He had no more reasons for this conception than for the geocentric theory, excepting that it violated no laws or principles, was in harmony with the known facts, and was simpler.

Contemporaries and successors of Copernicus were far from unanimous in accepting the heliocentric theory. One of the dissenters of the succeeding generation is worthy of note for his logical though erroneous argument against it. Tycho Brahe* contended that the Copernican theory was impossible, because if the earth revolved around the sun, and at one season was at one side of its orbit, and at another was on the opposite side, the stars would apparently change their positions in relation to the earth (technically, there would be an annual parallax), and he could detect no such change. His reasoning was perfectly sound, but was based upon an erroneous conception of the distances of the stars. The powerful instruments of the past fifty years have made these parallactic motions of many of the stars a determinable, though a very minute, angle, and constitute an excellent proof of the heliocentric theory (see p. 110).

Nine years after the death of Brahe, Galileo Galilei (1564– 1642) by the use of his recently invented telescope discovered that there were moons revolving about Jupiter, indicating by analogy the truth of the Copernican theory. Following upon the heels of this came his discovery that Venus in its swing back and forth near the sun plainly shows phases just as our moon does, and appears larger when in the crescent than when in the full. The only logical conclusion was that it revolves around the sun, again confirming by analogy the Copernican theory. Galilei was a thorough-going Copernican in private belief, but was not permitted to teach the doctrine, as it was considered unscriptural.

As an illustration of the humiliating subterfuges to which he was compelled to resort in order to present an argument based upon the heretical theory, the following is a quotation from an argument he entered into concerning three comets which appeared in 1618. He based his argument as to their

*Tycho Brahe (1546–1601) a famous Swedish astronomer, was born at Knudstrup, near Lund, in the south of Sweden, but spent most of his life in Denmark.

motions upon the Copernican system, professing to repudiate that theory at the same time.

"Since the motion attributed to the earth, which I as a pious and Christian person consider most false, and not to exist, accommodates itself so well to explain so many and such different phenomena, I shall not feel sure that, false as it is, it may not just as deludingly correspond with the phenomena of comets."

One of the best supporters of this theory in the next generation was Kepler (1571–1630), the German astronomer, and friend and successor of Brahe. His laws of planetary motion (see p. 283)were, of course, based upon the Copernican theory, and led to Newton's discovery of the laws of gravitation.

James Bradley (1693–1762) discovered in 1727 the aberration of light (see p. 105), and the supporters of the Ptolemaic system were routed, logically, though more than a century had to pass before the heliocentric theory became universally accepted.

APPENDIX

GRAVITY

Gravity is frequently defined as the earth's attractive influence for an object. Since the attractive influence of the mass of the earth for an object on or near its surface is lessened by centrifugal force (see p. 14) and in other ways (see p. 183), it is more accurate to say that the force of gravity is the resultant of

a. The attractive force mutually existing between the earth and the object, and

b. The lessening influence of centrifugal force due to the earth's rotation.

Let us consider these two factors separately, bearing in mind the laws of gravitation (see p. 16).

a. Every particle of matter attracts every other particle.

(1) Hence the point of gravity for any given object on the surface of the earth is determined by the mass of the object itself as well as the mass of the earth. The object pulls the earth as truly and as much as the earth attracts the object. The common center of gravity of the earth and this object lies somewhere between the center of the earth's mass and the center of the mass of the object. Each object on the earth's surface, then, must have its own independent common center of gravity between it and the center of the earth's mass. The position of this common center will vary—

(*a*) As the object varies in amount of matter (first law), and

(*b*) As the distance of the object from the center of the earth's mass varies (inversely as the square of the distance).

(2) Because of this principle, the position of the sun or moon slightly modifies the exact position of the center of gravity just explained. It was shown in the discussion of tides that, although the tidal lessening of the weight of an object is as yet an immeasurable quantity, it is a calculable one and produces tides (see p. 183).

b. The rotation of the earth gives a centrifugal force to every object on its surface, save at the poles.

(1) Centrifugal force thus exerts a slight lifting influence on objects, increasing toward the equator. This lightening influence is sufficient to decrease the weight of an object at the equator by $\frac{1}{289}$ of the whole. That is to say, an object which weighs 288 pounds at the equator would weigh a pound more if the earth did not rotate. Do not infer from this that the centrifugal force at the pole being zero, a body weighing 288 pounds at the equator would weigh 289 pounds at the pole, not being lightened by centrifugal force. This would be true *if the earth were a sphere*. The bulging at the equator decreases a body's weight there by $\frac{1}{599}$ as compared with the weight at the poles. Thus a body at the equator has its weight lessened by $\frac{1}{289}$ because of rotation and by $\frac{1}{599}$ because of greater distance from the center, or a total of $\frac{1}{195}$ of its weight as compared with its weight at the pole. A body weighing 195 pounds at the pole, therefore, weighs but 194 pounds at the equator. Manifestly the rate of the earth's rotation determines the amount of this centrifugal force. If the earth rotated seventeen times as fast, this force at the equator would exactly equal the earth's attraction,* objects there

*Other things equal, centrifugal force varies with the square of the velocity (see p. 13), and since centrifugal force at the equator equals 289

would have no weight; that is, gravity would be zero. In such a case the plumb line at all latitudes would point directly toward the nearest celestial pole. A clock at the 45th parallel with a pendulum beating seconds would gain one beat every 19½ minutes if the earth were at rest, but would lose three beats in the same time if the earth rotated twice as fast.

(2) Centrifugal force due to the rotation of the earth not only a"ects the amount of gravity, but modifies the direction in which it is exerted. Centrifugal force acts in a direction at right angles to the axis, not directly opposite the earth's attraction excepting at the equator. Thus plumb lines, excepting at the equator and poles, are slightly tilted toward the poles.

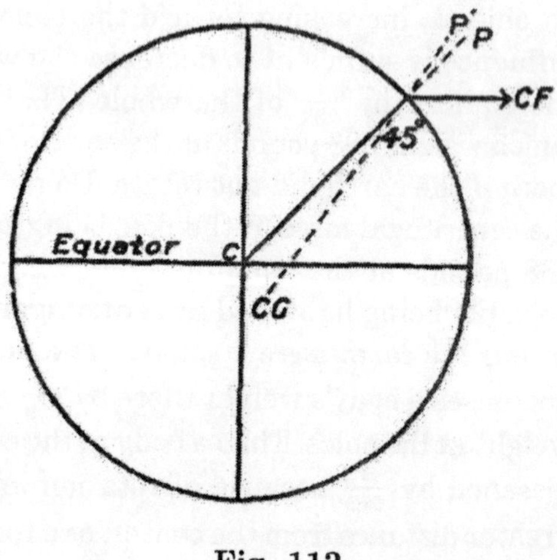

Fig. 112

If the earth were at rest a plumb line at latitude 45° would be in the direction toward the center of the mass of the earth at *C* (Fig. 112). The plumb line would then be *PC*. But centrifugal force is exerted toward *CF*, and the resultant of the attraction toward *C* and centrifugal force toward *CF* makes the line deviate to a point between those direc-tions, as *CG*, the true center of gravity, and the plumb line becomes *P′CG*. The amount of the centrifugal force is so small as compared with the earth's attraction that this deviation is not great. It is greatest at the 45th parallel where it amounts to 5′ 57″, or nearly one tenth of a degree.

There is an almost equal deviation due to the oblateness of the earth. At latitude 45. the total deviation of the plumb line from a line drawn to the center of the earth is 11° 30′65″.

LATITUDE

In Chapter II the latitude of a place was simply defined as the arc of a meridian intercepted between that place and the equator. This is true geographical latitude, but the discussion of *gravity* places us in a position to understand astronomical and geocentric latitude, and how geographic latitude is determined from astronomical latitude.

Owing to the elliptical form of a meridian "circle," the vertex of the angle constituting the latitude of a place is not at the center of the globe. A portion of a meridian circle near the equator is an arc of a smaller circle than a portion of the same meridian near the pole (see p. 41 and Fig. 18).

Geocentric Latitude. It is sometimes of value to speak of the angle formed at the center of the earth by two lines, one drawn to the place whose latitude is sought, and the other to the equator on the same meridian. This is called the geocentric latitude of the place.

Astronomical Latitude. The astronomer ascertains latitude from celestial measurements by reference to a level line or a plumb line. Astronomical latitude, then, is the angle formed between the plumb line and the plane of the equator.

In the discussion of gravity, the last effect of centrifugal force noted was on the direction of the plumb line. It was shown that this line, excepting at the equator and poles, is deviated slightly toward the pole. The effect of this is to increase correspondingly the astronomical latitude of a place. Thus at latitude 45° , astronomical latitude is increased by 5′ 57″, the amount of this deviation. If there were no rotation of the earth, there would be no deviation of the plumb line, and what we call latitude 60° would become 59° 54.′51″. Were the earth to rotate twice as fast, this latitude, as determined by the same astronomical instruments, would become 60° 15′ 27″.

If adjacent to a mountain, the plumb line deviates toward the mountain because of its attractive inuence on the plumb bob; and other deviations are also observed, such as with the ebb and flow of a near by tidal wave. These deviations are called "station errors," and allowance must be made for them in making all calculations based upon the plumb line.

Geographical latitude is simply the astronomical latitude, *corrected* for the deviation of the plumb line. Were it not for these deviations the latitude of a place would be determined within a few feet of perfect accuracy. As it is, errors of a few hundred feet sometimes may occur (see p. 287).

Celestial Latitude. In the discussion of the celestial sphere many circles of the celestial sphere were described in the same terms as circles of the earth. The celestial equator, Tropic of Cancer, etc., are imaginary circles which correspond to the terrestrial equator, Tropic of Cancer, etc. Now as terrestrial latitude is distance in degrees of a meridian north or south of the equator of the earth, one would infer that celestial latitude is the corresponding distance along a celestial meridian from the celestial equator, but this *is not the case.* Astronomers reckon celestial latitude from the *ecliptic* instead of from the celestial equator. As previously explained, the distance in degrees from the celestial equator is called *declination.*

Celestial Longitude is measured in degrees along the ecliptic from the vernal equinox as the initial point, measured always eastward the 360° of the ecliptic.

In addition to the celestial pole 90° from the celestial equator, there is a pole of the ecliptic, 90° from the ecliptic. A celestial body is thus located by reference to two sets of circles and two poles.

(a) Its declination from the celestial equator and position in relation to hour circles, as celestial meridians are commonly called (see Glossary).

(b) Its celestial latitude from the ecliptic and celestial longitude from "ecliptic meridians."

KEPLER'S LAWS

These three laws find their explanation in the laws of gravitation, although Kepler discovered them before Newton made the discovery which has immortalized his name.

First Law. The orbit of each planet is an ellipse, having the sun as a focus.

Second Law. The planet moves about the sun at such rates that the straight line connecting the center of the sun with the center of the planet (this line is called the planet's radius vector) sweeps over equal areas in equal times (see Fig. 113).

The distance of the earth's journey for each of the twelve months is such that the ellipse is divided into twelve equal areas. In the discussion of seasons we observed (p. 169) that when in perihelion, in January, the earth receives more heat each day than it does when in aphelion, in July. The northern hemisphere, being turned away from the sun in January, thus has warmer winters than it would otherwise have, and being toward the sun in July, has cooler summers. This is true only for corresponding days, not for the seasons as a whole. According to Kepler's second law the earth must receive exactly the same *total amount* of heat from the vernal equinox (March 21) to the autumnal equinox (Sept. 23), when farther from the sun, as from the autumnal to the vernal equinox, when nearer the sun. During the former period, the northern summer, the earth receives less heat day by day, but there are more days.

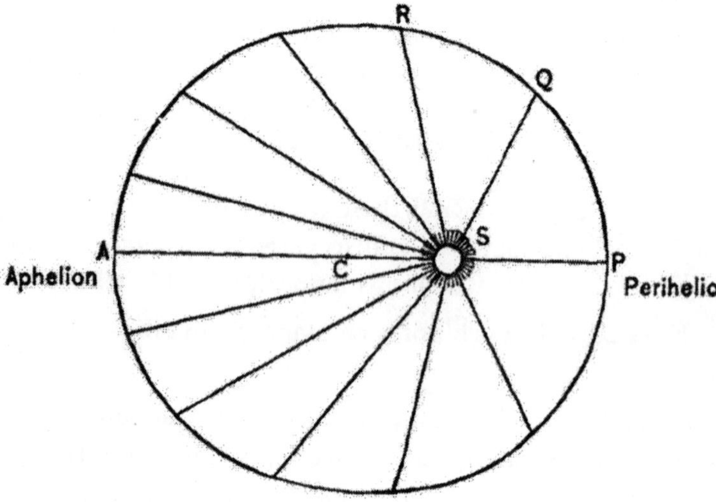

Fig. 113

Third Law. The squares of the lengths of the times (sidereal years) of planets are proportional to the cubes of their distances from the sun. Thus,

$$(\text{Earth's year})^2 : (\text{Mars' year})^2 :: (\text{Earth's distance})^3 : (\text{Mars' distance})^3 .$$

Knowing the distance of the earth to the sun and the distance of a planet to the sun, we have three of the quantities for our proportion, calling the earth's year 1, and can find the year of the planet; or, knowing the time of the planet, we can find its distance.

MOTIONS OF THE EARTH'S AXIS

In the chapter on seasons it was stated that excepting for exceedingly slow or minute changes the earth's axis at one time is parallel to itself at other times. There are three such motions of the axis.

Precession of the Equinoxes. Since the earth is slightly oblate and the bulging equator is tipped at an angle of 23½° to the ecliptic, the sun's attraction on this rim tends to draw the axis over at right angles to the equator. The rotation of the earth, however, tends to keep the axis parallel to itself, and the effect of the additional acceleration of the equator is to cause the axis to rotate slowly, keeping the same angle to the ecliptic, however.

At the time of Hipparchus (see p. 274), who discovered this rotation of the axis, the present North star, Alpha Ursa Minoris, was about 12° from the true pole of the celestial sphere, toward which the axis points. The course which the pole is taking is bringing it somewhat nearer the polestar; it is now about 1° 15′ away, but a hundred years hence will be only half a degree from it. The period of this rotation is very long, about 25, 000 years, or 50.2′ each year. Ninety degrees from the ecliptic is the pole of the ecliptic about which the pole of the celestial equator rotates, and from which it is distant 23½°.

As the axis rotates about the pole of the ecliptic, the point where the plane of the equator intersects the plane of the ecliptic, that is, the equinox, gradually shifts around westward. Since the vernal equinox is at a given point in the earth's orbit one year, and the next year is reached a little ahead of where it was the year before, the term *precession of the equinoxes* is appropriate. The sidereal year (see p. 133) is the time required for the earth to make a complete revolution in its orbit. A solar or tropical year is the interval from one vernal equinox to the next vernal equinox, and since the equinoxes "precede," a tropical year ends about twenty minutes before the earth reaches the same point in its orbit a second time.

As is shown in the discussion of the earth's revolution (p. 169),the earth is in perihelion December 31, making the northern summer longer and cooler, day by day, than it would otherwise be, and the winter shorter and warmer. The traveling of the vernal equinox around the orbit, however, is gradually shifting the date of perihelion, so that in ages yet to come perihelion will be reached in July, and thus terrestrial climate is gradually changing. This perihelion point (and with it, aphelion) has a slight westward motion of its own of 11.25′. each year, making, with the addition of the precession of the equinoxes of 50.2′, a total shifting of the perihelion point (see "Apsides"in the Glossary) of 1′ 1.45″. At this slow rate, 10,545 years must pass before perihelion will be reached July 1. The amount of the ellipticity of the earth's orbit is gradually decreasing, so that by the time this shifting has taken place the orbit will be so nearly circular that there may be but slight climatic effects of this shift of perihelion. It may be of interest to note that some have reasoned that ages ago the earth's orbit was so elliptical that the northern winter, occurring in aphelion, was so long and cold that great glaciers were formed in northern North America and Europe which the short, hot summers could not melt. The fact of the glacial age cannot be disputed, but this explanation is not generally accepted as satisfactory.

Nutation of the Poles. Several sets of gravitative influences cause a slight periodic motion of the earth's axis toward and from the pole of the ecliptic. Instead of "preceding" around the circle 47° in diameter, the axis makes a slight wavelike motion, a "nodding," as it is called. The principal nutatory motion of the axis is due to the fact that the moon's orbit about the earth (inclined 5° 8′ to the ecliptic) glides about the ecliptic in 18 years, 220 days, just as the earth's equator glides about the ecliptic once in 25,800 years. Thus through periods of nearly nineteen years each the obliquity of the ecliptic (see pp. 119, 148) gradually increases and decreases again. The rate of this nutation varies somewhat and is always very slight; at present it is 0.47″ in a year.

Wandering of the Poles. In the discussion of gravity (p. 278), it was shown that any change in the position of particles of matter effects a change in the point of gravity common to them. Slight changes in the crust of the earth are constantly taking place, not simply the gradational changes of wearing down mountains and building up of depositional features, but great diastrophic changes in mountain structure and continental changes of level. Besides these physiographic changes, meteorological conditions must be factors in displacement of masses, the accumulation of snow, the uctuation in the level of great rivers, etc. For these reasons minute changes in the position of the axis of rotation must take place within the earth. Since 1890 such changes in the position of the axis within the globe have been observed and recorded. The "wandering of the poles," as this slight shifting of the axis is called, has been demonstrated by the variation in the latitudes of places. A slight increase in the latitude of an observatory is noticed, and at the same time a corresponding decrease is observed in the latitude of an observatory on the opposite side of the globe. "So definite are the processes of practical astronomy that the position of the north pole can be located with no greater uncertainty than the area of a large Eskimo hut."*

*Todd's New Astronomy, p. 95.

In 1899 the International Geodetic Association took steps looking to systematic and careful observations and records of this wandering of the poles. Four stations not far from the thirty-ninth parallel but widely separated in longitude were selected, two in the United States, one in Sicily, and the other in Japan.

All of the variations since 1889 have been within an area less than sixty feet in diameter.

Seven Motions of the Earth. Seven of the well-defined motions of the earth have been described in this book:

1. Diurnal Rotation.

2. Annual Revolution in relation to the sun.

3. Monthly Revolution in relation to the moon (see p. 184).

4. Precessional Rotation of Axis about the pole of the ecliptic.

5. Nutation of the poles, an elliptical or wavelike motion in the precessional orbit of the axis.

6. Shifting on one axis of rotation, then on another, leading to a "wandering of the poles."

7. Onward motion with the whole solar system (see "Sun's Onward Motion," p. 248).

MATHEMATICAL TREATMENT OF TIDES

The explanation of the cause of tides in the chapter on that subject may be relied upon in every particular, although mathematical details are omitted. The mathematical treatment is difficult to make plain to those who have not studied higher mathematics and physics. Simplified as much as possible, it is as follows:

Let it be borne in mind that to find the cause of tides we must find *unbalanced forces which change their positions.*

Surface gravity over the globe varies slightly in different places, being less at the equator and greater toward the poles. As shown elsewhere, the force of gravity at the equator is less for two reasons:

a. Because of greater centrifugal force.

b. Because of the oblateness of the earth.

(a) Centrifugal force being greater at the equator than elsewhere, there is an unbalanced force which must cause the waters to pile up to some extent in the equatorial region. If centrifugal force were sometimes greater at the equator and sometimes at the poles, there would be a corresponding shifting of the accumulated waters and we should have a tide—and it would be an immense one. But we know that this unbalanced force does not change its position, and hence it cannot produce a tide.

(b) Exactly the same course of reasoning applies to the unbalanced force of gravity at the equator due to its greater distance from the center of gravity. The position of this unbalanced force does not shift, and no tide results.

Since the earth turns on its axis under the sun and moon, any unbalanced forces they may produce will necessarily shift as different portions of the earth are successively turned toward or from them. Our problem, then, is to find the cause and direction of the unbalanced forces produced by the moon or sun.

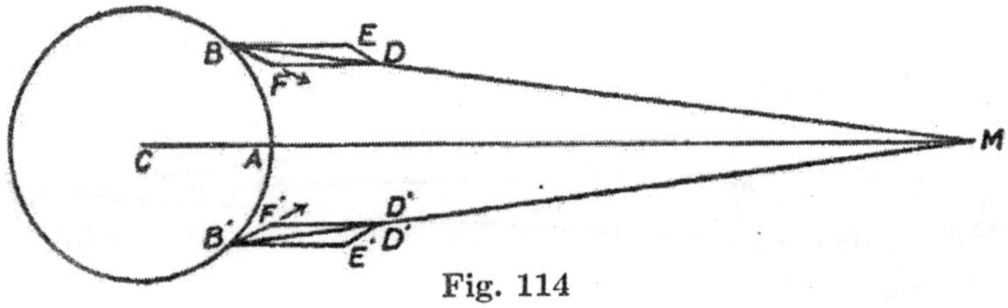

Fig. 114

In Figure 114, let CA be the acceleration toward the moon at *C*, due to the moon's attraction. Let *BD* be the acceleration at *B*. Now *B* is nearer the moon than *C*, so *BD* will be greater than *CA*, since the attraction varies inversely as the square of the distance.

From *B* construct *BE* equal to *CA*. Comparing forces *BE* and *BD*, the latter is greater. Completing the parallelogram, we have *BFDE*. Now it is a simple demonstration in physics that if two forces act upon *B*, one to *F* and the other to *E*, the resultant of the two forces will be the diagonal *BD*. Since *BE* and *BF* combined result in *BD*, it follows that *BF* represents the unbalanced force at *B*.

At *B*, then, there is an unbalanced force as compared with C as represented by BF. At B′ the unbalanced force is represented by B′F′. Note the pulling direction in which these unbalanced forces are exerted.

Note.—For purposes of illustration the distance of the moon represented in the figures is greatly diminished. The distance *CA* is taken arbitrarily, likewise the distance *BD*. If *CA* were longer, however, *BD* would be still longer; and while giving *CA* a different length, would modify the form of the diagram, the mathematical relations would remain unchanged. Because of the short distance given *CM* in the figures, the difference between the *BF* in Figure 114 and *BF* in Figure 115 is greatly exaggerated. The difference between the unbalanced or tide-producing force on the side toward the moon and that on the opposite side is approximately .0467 *BF* (Fig. 114).

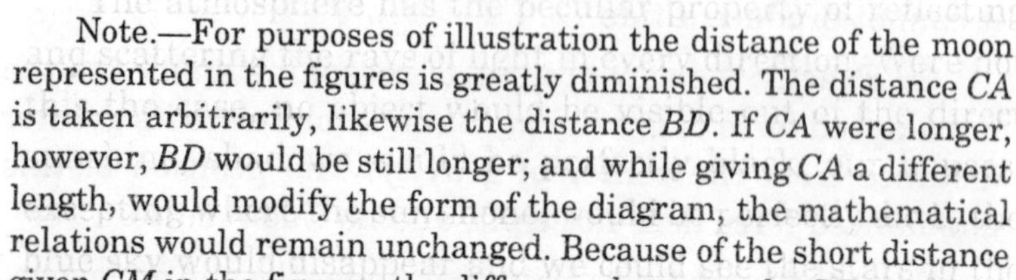

Fig. 115

In Figure 115, *B* is farther from the moon than *C*, hence BE (equal to *CA*) is greater than *BD*, and the unbalanced force at *B* is *BF*, directed away from the moon. A study of Figures 114 and 115 will show that the unbalanced force on the side towards the moon (*BF* in Fig. 114) is slightly greater than the unbalanced force on the side opposite the moon (*BF* in Fig.115).The difference, however, is exceedingly slight, and the tide on the opposite side is practically equal to the tide on the side toward the attracting body.

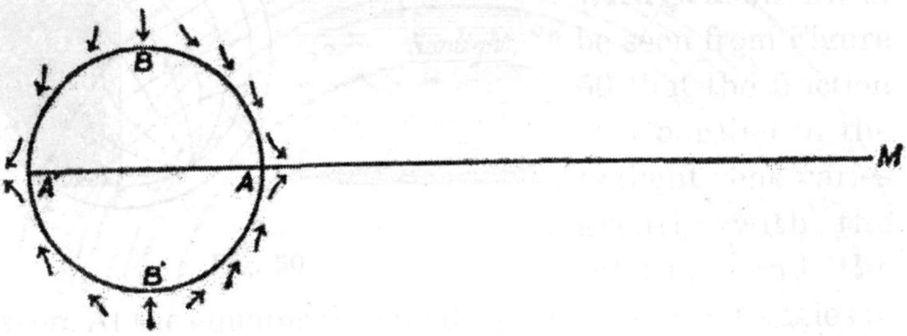

Fig. 116

Combining the arrows showing the directions of the unbalanced forces in the two figures, we have the arrows shown in Figure 116. The distribution and direction of the unbalanced forces may be thus summarized: "The disturbing force produces a pull along *AA´* and a squeeze along *BB´*."*

THE ZODIAC

This belt in the celestial sphere is 16° wide with the ecliptic as the center. The width is purely arbitrary. It could have been wider or narrower just as well, but was adopted by the ancients because the sun, moon, and planets known

*Mathematical Astronomy, Barlow and Bryan, p. 377.

to them were always seen within 8° of the pathway of the sun. We know now that several asteroids, as truly planets as the earth, are considerably farther from the ecliptic than 8° indeed, Pallas is sometimes 34. from the ecliptic—to the north of overhead to people of northern United States or central Europe.

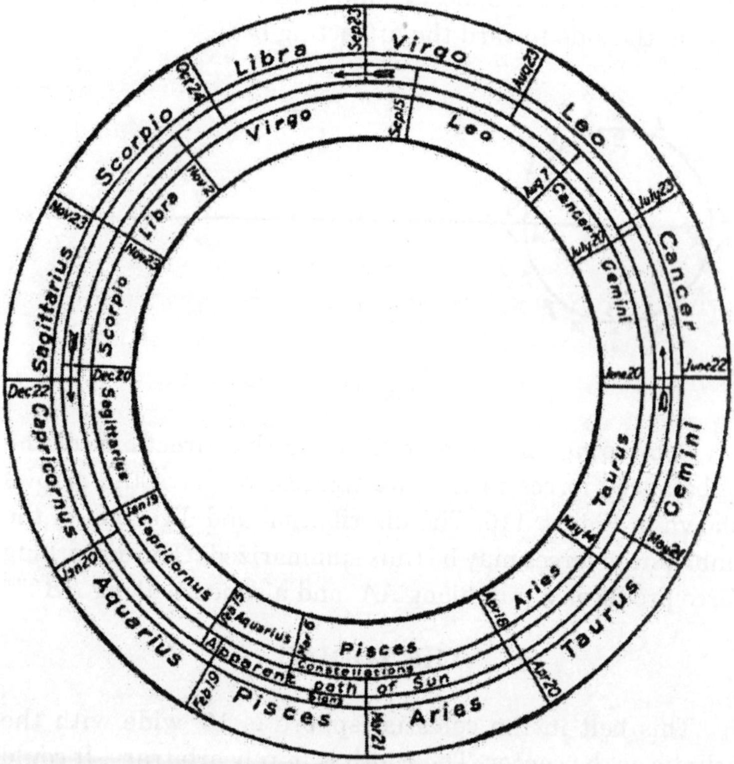

Fig. 117

Signs. As the sun "creeps backward" in the center of the zodiac, one revolution each year, the ancients divided its pathway into twelve parts, one for each month. To each of these sections of thirty degrees ($360^\circ \div 12 = 30^\circ$) names were assigned, all but one after animals, each one being considered appropriate as a "sign" of an annual recurrence (see p. 118). Aries seems commonly to have been taken as the first in the series, the beginning of spring. Even yet the astronomer counts the tropical year from the "First point of Aries," the moment the center of the sun crosses the celestial equator on its journey northward.

As explained in the discussion of the precession of the equinoxes (p. 285), the point in the celestial equator where the center of the sun crosses it shifts westward one degree in about seventy years. In ancient days the First point of Aries was in the constellation of that name but now it is in the constellation to the west, Pisces. The sign Aries begins with the First point of Aries, and thus with the westward travel of this point all the signs have moved back into a constellation of a different name. Another difference between the signs and the constellations of the zodiac is that the star clusters are of unequal length, some more than 30° and some less, whereas the signs are of uniform length. The positions and widths of the signs and constellations with the date when the sun enters each are shown in Figure 117.

Aries, the first sign, was named after the ram, probably because to the ancient Chaldeans, where the name seems to have originated, this was the month of sacrifice. The sun is in Aries from March 21 until April 20. It is represented by a small picture of a ram  or by a hieroglyphic (♈).

Taurus, the second sign (), was dedicated to the bull. In ancient times this was the first of the signs, the vernal equinox being at the beginning of this sign. According to very ancient mythology it was the bull that drew the sun along its "furrow" in the sky. There are, however, many other theories as to the origin of the designation. The sun is in Taurus from April 20 until May 21.

Gemini, the third sign, signifies twins () and gets its name from two bright stars, Castor and Pollux, which used to be in this sign, but are now in the sign Cancer. The sun is in Gemini from May21 until June22.

Cancer, the fourth sign (), was named after the crab, probably from the fact that when in this sign the sun retreats back again, crablike, toward the south. The sun is in Cancer from June 22 until July 23.

Leo, signifying lion, is the fifth sign () and seems to have been adopted because the lion usually was used as a symbol for fire, and when the sun was in Leo the hottest weather occurred. The sun is in this sign from July 23 until August 23.

Virgo, the virgin (), refers to the Chaldean myth of the descent of Ishtar into hades in search of her husband. The sun is in Virgo from August 23 until September 23.

The foregoing are the summer signs and, consequently, the corresponding constellations are our winter constellations. It must be remembered that the sign is always about 30° (the extreme length of the "Dipper") to the west of the constellation of the same name.

Libra, the balances (), appropriately got its name from the fact that the autumnal equinox, or equal balancing of day and night, occurred when the sun was in the constellation thus named the Balances. The sun is now in Libra from September 23 until October 24.

Scorpio is the eighth sign (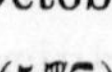). The scorpion was a symbol of darkness, and was probably used to represent the shortening of days and lengthening of nights. The sun is now in Scorpio from October 24 until November 23.

Sagittarius, meaning an archer or bowman, is sometimes represented as a Centaur with a bow

and arrow. The sun is in this sign from November 23 until December 22.

Capricorn, signifying goat, is often represented as having the tail of a fish (). It probably has its origin as the mythological nurse of the young solar god. The sun is in Capricorn from December22 until January20.

Aquarius, the water-bearer (), is the eleventh sign and probably has a meteorological origin, being associated as the cause of the winter rains of Mediterranean countries. The sun is in this sign from January 20 until February 19.

Pisces is the last of the twelve signs. In accordance with the meaning of the term, it is represented as two fishes (). Its significance was probably the same as the water-bearer. The sun is in this sign from February 19 until the vernal equinox, March 21, when it has completed the "labors" of its circuit, only to begin over again.

The twelve signs of the ancient Chinese zodiac were dedicated to a quite different set of animals; being, in order, the Rat, the Ox, the Tiger, the Hare, the Dragon, the Serpent, the Horse, the Sheep, the Monkey, the Hen, the Dog, and the Pig. The Egyptians adopted with a few changes the signs of the Greeks.

MYTHS AND SUPERSTITIONS AS TO THE RELATION OF THE ZODIAC TO THE EARTH

When one looks at the wonders of the heavens it does not seem at all strange that in the early dawn of history, ignorance and superstition should clothe the mysterious luminaries of the sky with occult influences upon the earth, the weather, and upon human a"airs. The ancients, observing the apparent fixity of all the stars excepting the seven changing ones of the zodiac—the sun, moon, and five planets known to them—endowed this belt and its seven presiding deities with special guardianship of the earth, giving us seasons, with varying length of day and change of weather; bringing forth at its will the sprouting of plants and fruitage and harvest in their season; counting off inevitably the years that span human life; bringing days of prosperity to some and of adversity to others; and marking the wars and struggles, the growth and decay of nations. With such a background of belief, at once their science and their religion, it is not strange that when a child was born the parents hastened to the astrologer to learn what planet or star was in the ascendancy, that is, most prominent during the night, and thus learn in advance what his destiny would be as determined from the character of the star that would rule his life.

The moon in its monthly path around the earth must pass through the twelve signs of the zodiac in 29½ days or spend about 2½ days to each sign. During the blight of intelligence ofthe dark ages, some medi val astrologer conceived the simple method of subdividing the human body into twelve parts to correspond to the twelve constellations of the zodiac. Beginning with the sign Aries, he dedicated that to the head, the neck he assigned to Taurus, the arms were given over to Gemini, the stars of Cancer were to rule the breast, the heart was presided over by Leo, and so on down to Pisces which was to rule the feet. Now anyone who was born when the moon was in Aries would be strong in the head, intellectual; if in Taurus, he would be strong in the neck and self-willed, etc. Moreover, since the moon makes a circuit of the signs of the zodiac in a month, according to his simple scheme when the moon is in Aries the head is especially affected; then diseases of the head rage (or is it then that the head is stronger to resist disease?), and during the next few days when the moon is in Taurus, beware of affections of the neck, and so on down the list. The very simplicity of this scheme and ease by which it could be remembered led to its speedy adoption by the masses who from time immemorial have sought explanations of various phenomena by reference to celestial bodies.

Now there is no astronomical or geographical necessity for considering Aries as the first sign of the zodiac. Our year begins practically with the advent of the sun into Capricorn—the beginning of the year was made January 1 for this very purpose. The moon is not in any peculiar position in relation to the earth March 21 any more than it is December 23. If when the calendar was revised the numbering of the signs of the zodiac had been changed also, then Capricorn, the divinities of which now rule the *knees,* would have been made to rule the *head*, and the whole artificial scheme would have been changed! Besides, the sign Capricorn does not include the *constellation* Capricorn, so with the precession of the equinoxes the subtle influences once assigned to the heavenly bodies of one constellation have been shifted to an entirely different set of stars! The association of storms with the sun's crossing the equinox and with the angle the cusps of the moon show to the observer (a purely geometric position varying with the position of the observer) is in the same class as bad luck attending the taking up of the ashes after the sun has gone down or the wearing of charms against rheumatism or the "evil" eye.

> "The fault, dear Brutus, is not in our stars,
> But in ourselves, that we are underlings."
>
> —Shakespeare.

PRACTICAL WORK IN MATHEMATICAL GEOGRAPHY

Concrete work in this subject has been suggested directly, by implication, or by suggestive queries and problems throughout the book. No instruments of specific character have been suggested for use excepting such as arè easily provided, as a graduated quadrant, compasses, an isosceles right triangle, etc. Interest in the subject will be greatly augmented if the following simple instruments, or similar devices, are made or purchased and used.

TO MAKE A SUNDIAL

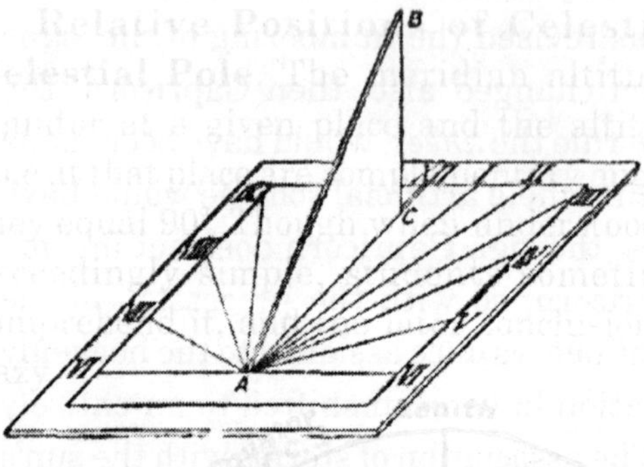

Fig. 118

This is not especially difficult and may be accomplished in several ways. A simple plan is shown in Figure 118.Angle BAC should be the co-latitude of the place, that is, the latitude subtracted from 90°, though this is not at all essential. The hour lines may be marked off according to two systems, for standard time or for local time.

Standard Time Dial. If you wish your dial to indicate clock time as correctly as possible, it will be necessary to consult the analemma or an almanac to ascertain the equation of time when the hour lines are drawn. Since the sun is neither fast nor slow April 14, June 15, September 1, or December 25, those are the easiest days on which to lay off the hours. On one of those dates you can lay them off according to a reliable timepiece.

If you mark the hour lines at any other date; ascertain the equation of time (see p. 128) and make allowances accordingly. Suppose the date is October 27. The analemma shows the sun to be 16 minutes fast. You should mark the hour lines that many minutes before the hour as indicated by your timepiece, that is, the noon line when your watch says 11:44 o'clock, the 1 o'clock line when the watch indicates 12:44, etc. If the equation is slow, say five minutes, add that time to your clock time, marking the noon line when your watch indicates 12:05, the next hour line at 1:05, etc. It is well to begin at the hour for solar noon, at that time placing the board so that the sun's shadow is on the XII mark and after marking off the afternoon hours measure from the XII mark westward corresponding distances for the forenoon. Unless you chance to live upon the meridian which gives standard time to the belt in which you are, the noon line will be somewhat to the east or west of north.

This sundial will record the apparent solar time of the meridian upon which the clock time is based. The difference in the time indicated by the sundial and your watch at any time is the equation of time. Test the accuracy of your sundial by noticing the time by your watch when the sundial indicates noon and comparing this difference with the equation of time for that day. If your sundial is accurate, you can set your watch any clear day by looking up the equation of time and making allowances accordingly. Thus the analemma shows that on May 28 the sun is three minutes fast. When the sundial indicates noon you know it is three minutes before twelve by the clock.

Local Time Dial. To mark the hour lines which show the local mean solar time (see p. 63), set the XII hour line due north. Note accurately the clock time when the shadow is north. One hour later mark the shadow line for the I hour line, two hours later mark the II hour line, etc. This dial will indicate the apparent solar time of your meridian. You can set your watch by it by first converting it into mean solar time and then into standard time. (This is explained on p. 129)

It should be noted that these two sundials are exactly the same for persons who use local time, or, living on the standard time meridian, use standard time.

THE SUN BOARD

The uses of the mounted quadrant in determining latitude were shown in the chapter on seasons (see p. 172). Dr. J. Paul Goode, of the University of Chicago, has designed a very convenient little instrument which answers well for this and other purposes.

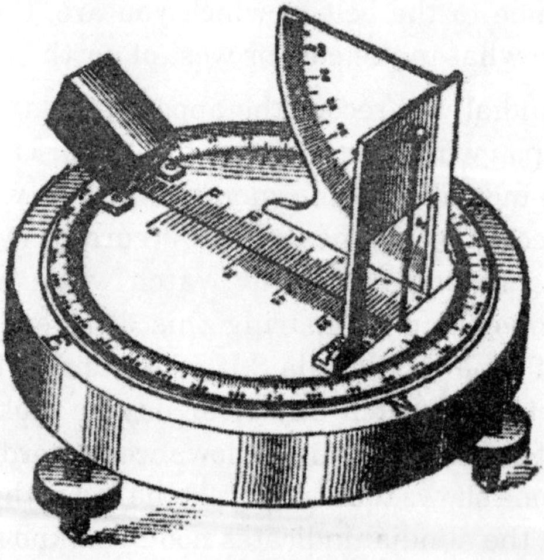

Fig. 119

A vertically placed quadrant enables one to ascertain the altitude of the sun for determining latitude and calculating the heights of objects. By means of a graduated circle placed horizontally the azimuth of the sun (see Glossary) may be ascertained. A simple vernier gives the azimuth readings to quarter degrees. It also has a device for showing the area covered by a sunbeam of a given size, and hence its heating power.

THE HELIODON

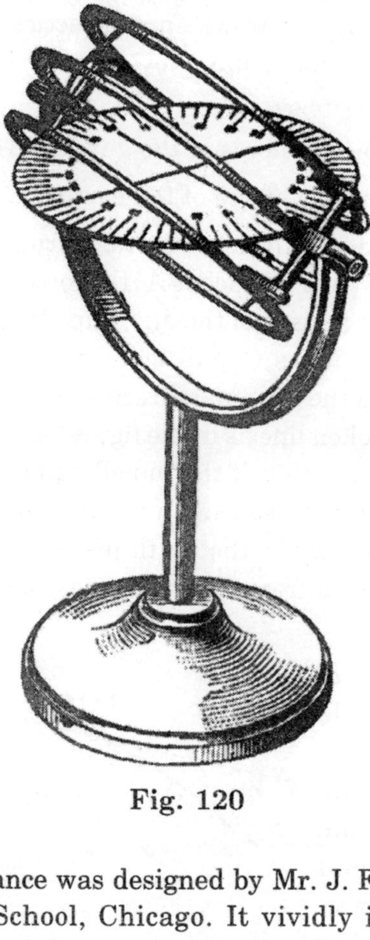

Fig. 120

This appliance was designed by Mr. J. F. Morse, of the Medill High School, Chicago. It vividly illustrates the apparent path of the sun at the equinoxes and solstices at any latitude. The points of sunrise and sunset can also be shown and hence the length of the longest day or night can be calculated.

WHAT KEEPS THE MEMBERS OF THE SOLAR SYSTEM

IN THEIR ORBITS?

When a body is thrown in a direction parallel to the horizon, as the bullet from a level gun, it is acted upon by two forces:

(a) The projectile force of the gun, *AB*. (Fig. 121.)

(b) The attractive force of the earth, *AC*.

The course it will actually take from point A is the diagonal *AA´*. When it reaches *A´* the force *AB* still acts (not considering the friction of the air), impelling it in the line *A´ B´*. Gravity continues to pull it in the line *A´ C´*, and the projectile takes the diagonal direction *A´A´´* and makes the curve (not a broken line as in the figure) *AA´A´´*. It is obvious from this diagram that if the impelling force be suffciently great, line *AB* will be so long in relation to line *AC* that the bullet will be drawn to the earth just enough to keep it at the same distance from the surface as that of its starting point.

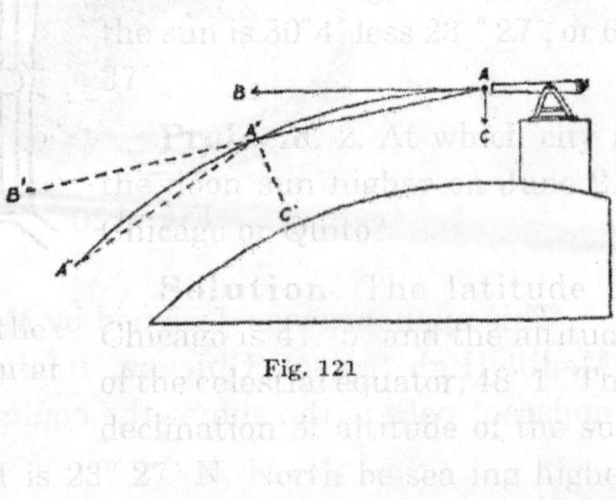

Fig. 121

The amount of such a projectile force near the surface of the earth at the equator as would thus keep an object at an unvarying distance from the earth is 26, 100 feet per second. Fired in a horizontal direction from a tower (not allowing for the friction of the air) such a bullet would forever circle around the earth. Dividing the circumference of the earth (in feet) by this number we find that such a bullet would return to its starting point in about 5,000 seconds, or 1 h. 23 m., making many revolutions around the earth during one day. Since our greatest guns, throwing a ton of steel a distance of twenty-one miles, give their projectiles a speed of only about 2, 600 feet per second, it will be seen that the rate we have given is a terrific one. If this speed were increased to 37,000 feet per second, the bullet would never return to the earth. One is tempted here to digress and demonstrate the utter impossibility of human beings even "making a trip to the moon," to say nothing of one to a much more distant planet. The terrific force with which we should have to be hurled to get away from the earth, fourteen times the speed of the swiftest cannon ball, is in itself an insuperable difficulty. Besides this, there would have to be the most exact calculation of the force and direction, allowing for *(a)* the curve given a projectile by gravity, *(b)* the centrifugal force of rotation, *(c)* the revolution of the earth, *(d)* the revolution of the moon, *(e)* the friction of the air, a variable quantity, impossible of calculation with absolute accuracy, *(f)* the inevitable swerving in the air by reason of its currents and varying density, and *(g)* the influence on the course by the attraction of the sun and planets. In addition to these mathematical calculations as to direction and projectile force, there would be the problem of *(h)* supply of air, *(i)* air pressure, to which our bodies through the evolution of ages have become adapted, *(j)* the momentum with which we would strike into the moon if we did "aim" right, etc.

Returning to our original problem, we may notice that if the bullet were fired horizontally at a distance of 4, 000 miles from the surface of the earth, the pull of gravity would be only one fourth as great (second law of gravitation), and the projectile would not need to take so terrific a speed to revolve around the earth. As we noticed in the discussion of Mars (see p. 255), the satellite Phobos is so near its primary, 1,600 miles from the surface, that it revolves at just about the rate of a cannon ball, making about three revolutions while the planet rotates once.

While allusion has been made only to a bullet or a moon, in noticing the application of the law of projectiles, the principle applies equally to the planets. Governed by the law here illustrated, a planet will revolve about its primary in an orbit varying from a circle to an elongated ellipse. Hence we conclude that a combination of projectile and attractive forces keeps the members of the solar system in their orbits.

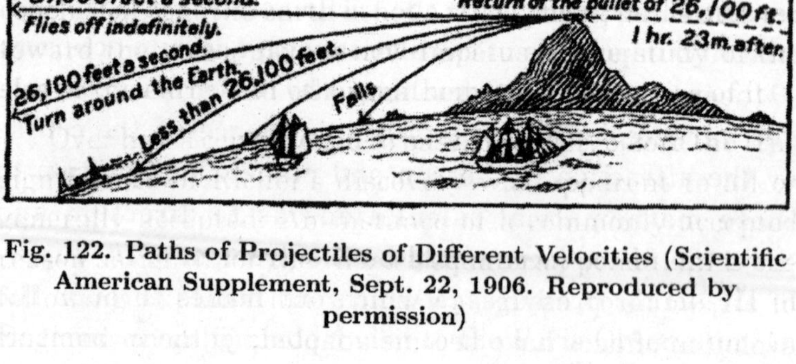

Fig. 122. Paths of Projectiles of Different Velocities (Scientific American Supplement, Sept. 22, 1906. Reproduced by permission)

FORMULAS AND TABLES

SYMBOLS COMMONLY EMPLOYED

There are several symbols which are generally used in works dealing with the earth, its orbit or some of its other properties. To the following brief list of these are added a few mathematical symbols employed in this book, which may not be familiar to many who will use it. The general plan of using arbitrary symbols is shown on page 13, where G represents universal gravitation and g represents gravity; C represents centrifugal force and c centrifugal force due to the rotation of the earth.

ϕ (Phi), latitude.

ε (Epsilon), obliquity of the ecliptic, also eccentricity of an ellipse.

π (Pi), the number which when multiplied by the diameter of a circle equals the circumference; it is 3.14159265, nearly 3.1416, nearly $3\frac{1}{7}$. $\pi^2 = 9.8696044$.

δ (Delta), declination, or distance in degrees from the celestial equator.

$\propto$, "varies as;" $x \propto y$ means x varies as y.

<, "is less than;" $x < y$ means x is less than y.

>, "is greater than;" $x > y$ means x is greater than y.

FORMULAS

The Circle and Sphere

r = radius. c = circumference.

d = diameter. a = area.

$\pi d = c$

$\frac{c}{\pi} = d$

πr^2 = area.

$4\pi r^2$ = surface of sphere.

$\frac{4}{3}\pi r^3$ = volume of sphere = 4.1888*r*3 (nearly).

The Ellipse

a = ½ major axis. o = oblateness.

b = ½ minor axis. e = eccentricity.

πab = area of ellipse.

$$o = \frac{a-b}{a}$$

$$e = \sqrt{\frac{a^2 - b^2}{a^2}}$$

The Earth Compared with Other Bodies

P = the radius of the body as compared with the radius of the earth. Thus in case of the moon, the moon's radius = 1081, the earth's radius = 3959, and $P = \frac{1081}{3959}$.

P^2 = surface of body as compared with that of the earth.

P^3 = volume of body as compared with that of the earth.

$\frac{mass}{P^2}$ = surface gravity as compared with that of the earth.

Centrifugal Force

c = centrifugal force. r = radius.

v = velocity. m = mass.

$$c = \frac{mv^2}{r}.$$

Lessening of surface gravity at any latitude by reason of the centrifugal force due to rotation.

g= surface gı avity.

c at any latitude $= \frac{g}{289} \times \cos^2 \phi$.

Deviation of the plumb line from true vertical by reason of centrifugal force due to rotation.

d = deviation.

$d = 357'' \times \sin 2\phi$.

Miscellaneous

Rate of swing of pendulum varies inversely as the square root of the surface gravity. $r = \frac{1}{\sqrt{g}}$.

Density of a body $= \frac{\text{mass}}{\text{vol.}}$.

Hourly deviation of the plane of a pendulum due to the rotation of the earth = sin latitude × 15° ($d = \sin\phi \times 15^\circ$).

Weight of bodies above the surface of the earth.

w = weight,

d = distance from the center of the earth.

$w \propto \frac{1}{d^2}$.

Weight of bodies below the surface of the earth. $w \propto d$.

GEOGRAPHICAL CONSTANTS*

Equatorial semi-axis:

in feet	20, 926, 062.
in meters	6, 378, 206.4
in miles	3, 963.307

*Dimensions of the earth are based upon the Clarke spheroid of 1866.

Polar semi-axis:	
in feet	20, 855, 121.
in meters	6, 356, 583.8
in miles	3, 949.871
Oblateness of earth	1 ÷ 294.9784
Circumference of equator (in miles)	24, 901.96
Circumference through poles (in miles)	24, 859.76
Area of earth's surface, square miles	196, 971, 984.
Volume of earth, cubic miles	259, 944, 035, 515.
Mean density (Harkness)	5.576
Surface density (Harkness)	2.56
Obliquity of ecliptic (see page 119)	23° 27′4. 98 s.
Sidereal year	365 d. 6 h. 9 m. 8.97 s. or 365.25636 d.
Tropical year	365 d. 5 h. 48 m. 45.51 s. or 365.24219 d.
Sidereal day	23 h. 56 m. 4.09 s. of mean solar time.
Distance of earth to sun, mean (in miles)	92, 800, 000.
Distance of earth to moon, mean (in miles)	238, 840.

MEASURES OF LENGTH

Statute mile	5, 280.00 feet
Nautical mile,* or knot	6, 080.27 "
German sea mile	6, 076.22 "
Prussian mile, law of 1868	24, 604.80 "
Norwegian and Swedish mile	36, 000.00 "
Danish mile	24, 712.51 "
Russian werst, or versta	3, 500.00 "
Meter	3.28 "
Fathom	6.00 "
Link of surveyor's chain	0.66 "

*As defined by the United States Coast and Geodetic Survey.

TABLE OF NATURAL SINES AND COSINES

Sin		Cos	Sin		Cos	Sin		Cos
0°	.0000	90°	31°	.5150	59°	61°	.8746	29°
1	.0175	89	32	.5299	58	62	.8829	28
2	.0349	88	33	.5446	57	63	.8910	27
3	.0523	87	34	.5592	56	64	.8988	26
4	.0698	86	35	.5736	55	65	.9063	25
5	.0872	85	36	.5878	54	66	.9135	24
6	.1045	84	37	.6018	53	67	.9205	23
7	.1219	83	38	.6157	52	68	.9272	22
8	.1392	82	39	.6293	51	69	.9336	21
9	.1564	81	40	.6424	50	70	.9397	20
10	.1736	80	41	.6561	49	71	.9455	19
11	.1908	79	42	.6691	48	72	.9511	18
12	.2079	78	43	.6820	47	73	.9563	17
13	.2250	77	44	.6947	46	74	.9613	16
14	.2419	76	45	.7071	45	75	.9659	15
15	.2588	75	46	.7193	44	76	.9703	14
16	.2756	74	47	.7314	43	77	.9744	13
17	.2924	73	48	.7431	42	78	.9781	12
18	.3090	72	49	.7547	41	79	.9816	11
19	.3256	71	50	.7660	40	80	.9848	10
20	.3420	70	51	.7771	39	81	.9877	9
21	.3584	69	52	.7880	38	82	.9903	8
22	.3746	68	53	.7986	37	83	.9925	7
23	.3907	67	54	.8090	36	84	.9945	6
24	.4067	66	55	.8192	35	85	.9962	5
25	.4226	65	56	.8290	34	86	.9976	4
26	.4384	64	57	.8387	33	87	.9986	3
27	.4540	63	58	.8480	32	88	.9994	2
28	.4695	62	59	.8572	31	89	.9998	1
29	.4848	61	60	.8660	30	90	1.0000	0
30	.5000	60						

TABLE OF NATURAL TANGENTS AND COTANGENTS

Tan		Cot	Tan		Cot	Tan		Cot
0°	.0000	90°	31°	.6009	59°	61°	1.8040	29°
1	.0175	89	32	.6249	58	62	1.8807	28
2	.0349	88	33	.6494	57	63	1.9626	27
3	.0524	87	34	.6745	56	64	2.0503	26
4	.0699	86	35	.7002	55	65	2.1445	25
5	.0875	85	36	.7265	54	66	2.2460	24
6	.1051	84	37	.7536	53	67	2.3559	23
7	.1228	83	38	.7813	52	68	2.4751	22
8	.1405	82	39	.8098	51	69	2.6051	21
9	.1584	81	40	.8391	50	70	2.7475	20
10	.1763	80	41	.8693	49	71	2.9042	19
11	.1944	79	42	.9004	48	72	3.0777	18
12	.2126	78	43	.9325	47	73	3.2709	17
13	.2309	77	44	.9657	46	74	3.4874	16
14	.2493	76	45	1.0000	45	75	3.7321	15
15	.2679	75	46	1.0355	44	76	4.0108	14
16	.2867	74	47	1.0724	43	77	4.3315	13
17	.3057	73	48	1.1106	42	78	4.7046	12
18	.3249	72	49	1.1504	41	79	5.1446	11
19	.3443	71	50	1.1918	40	80	5.6713	10
20	.3640	70	51	1.2349	39	81	6.1338	9
21	.3839	69	52	1.2794	38	82	7.1154	8
22	.4040	68	53	1.3270	37	83	8.1443	7
23	.4245	67	54	1.3764	36	84	9.5144	6
24	.4452	66	55	1.4281	35	85	11.43	5
25	.4663	65	56	1.4826	34	86	14.30	4
26	.4877	64	57	1.5399	33	87	19.08	3
27	.5095	63	58	1.6003	32	88	24.64	2
28	.5317	62	59	1.6643	31	89	57.29	1
29	.5543	61	60	1.7321	30	90	0.0000	0
30	.5774	60						

GLOSSARY

Aberration, the apparent displacement of sun, moon, planet, or star produced as a resultant of (*a*) the orbital velocity of the earth, and (*b*) the velocity of light from the heavenly body.

Acceleration, increase or excess of mean motion or velocity.

Altitude, elevation in degrees (or angle of elevation) of an object above the horizon.

Analemma, a scale showing (*a*) the mean equation of time and (*b*) the mean declination of the sun for each day of the year.

Aphelion (a fe´ li on), the point in a planet's orbit which is farthest from the sun.

Apogee (ap´ o je), the point farthest from the earth in any orbit; usually applied to the point in the moon's orbit farthest from the earth.

Apparent solar day, see Day.

Apparent (solar) time, see Time.

Apsides (ap´ si dez), line of, a line connecting perihelion and aphelion of a planet's orbit, or perigee and apogee of a moon's orbit. Apsides is plural for apsis, which means the point in an orbit nearest to the primary or farthest from it.

Arc, part of a circle; in geography, part of the circumference of a circle.

Asteroids, very small planets. A large number of asteroids revolve around the sun between the orbits of Mars and Jupiter.

Autumnal equinox, see Equinox.

Axis, the line about which an object rotates.

Azimuth (az´ i´ muth) the angular distance of an object from the celestial meridian of the place of the observer to the celestial meridian of the object. The azimuth of the sun is the distance in degrees from its point of rising or setting to a south point on the horizon.

Celestial sphere, the apparent hollow sphere in which the sun, moon, planets, comets, and stars seem to be located.

Center of gravity the point about which a body (or group of bodies) balances.

Centrifugal force (sen trif′ u gal), a force tending away from a center.

Centripetal force (sen trip′ e tal), a force tending toward a center.

Colures (kō lūrz′), the four principal meridians of the celestial sphere, two passing through the equinoxes and two through the solstices.

Conjunction, see Syzygy.

Copernican system (kō per′ ni can), the theory of the solar system advanced by Copernicus (1473–1543) that the sun is the center of the solar system, the planets rotating on their axes and revolving around the sun. See Heliocentric theory.

Co-tidal lines, lines passing through places that have high tide at the same time.

Day.

ASTRONOMICAL DAY, a period equal to a mean solar day, reckoned from noon and divided into twenty-four hours, usually numbered from one to twenty-four.

CIVIL DAY the same as an astronomical day excepting that it is reckoned from midnight. It is also divided into twenty-four hours, usually numbered in two series, from one to twelve.

SIDEREAL DAY, the interval between two successive passages of a celestial meridian over a given terrestrial meridian. The zero meridian from which the sidereal day is reckoned is the one passing through the First point of Aries. The length of the sidereal day is 23 h. 56 m. 4.09 s. The sidereal day is divided into twenty-four hours, each shorter than those of the civil or astronomical day; they are numbered from one to twenty-four.

SOLAR DAY

Apparent solar day, the interval between two successive passages of the sun's center over the meridian of a place; that is, from sun noon to the next sun noon; this varies in length from 23 h. 59 m. 38.8 s. to 24 h. 0 m. 30 s.

Mean solar day, the average interval between successive passages of the sun's center over the meridian of a place; that is, the average of the lengths of all the solar days of the year; this average is 24 h. as we commonly reckon civil or clock time.

Declination is the distance in degrees of a celestial body from the celestial equator. Declination in the celestial sphere corresponds to latitude on the earth.

Eccentricity (ek sen tris′ i ty), see Ellipse.

Ecliptic (e kl-p′ t-k), the path of the center of the sun in its apparent orbit in the celestial sphere. A great circle of the celestial sphere whose plane forms an angle of 23°27′ with the plane of the equator. This inclination of the plane of the ecliptic to the plane of the equator is called the *obliquity* of the ecliptic. The points 90° from the ecliptic are called the poles of the ecliptic. Celestial latitude is measured from the ecliptic.

Ellipse, a plane figure bounded by a curved line, every point of which is at such distances from two points within called the foci (pronounced fō′sī; singular, focus) that the sum of the distances is constant.

ECCENTRICITY (ek sen tris′ i ty) is the fraction obtained by dividing the distance of a focus to the center of the major axis by one half the major axis.

OBLATENESS or ellipticity is the deviation of an ellipse from a circle and is the fraction obtained by dividing the difference between the major and minor axes by the major axis.

Ellipticity (el l-p tis′ i ty), see Ellipse.

Equation of time (e kwā′ shun), the difference between apparent solar time, or time as actually indicated by the sun, and the mean solar time, or the average time indicated by the sun. It is usually indicated by the minus sign when the apparent sun is faster than the mean sun and with the plus sign when the apparent time is slow. The apparent sun time combined with the equation of time gives the mean time; *e.g.*, by the apparent sun it is 10 h. 30 m., the equation is 2 m. (sun fast 2 m.), combined we get 10 h. 28 m., the mean sun time. See Day.

Equator (e kwā′ ter), when not otherwise qualified means terrestrial equator.

CELESTIAL EQUATOR, the great circle of the celestial sphere in the plane of the earth's equator. Declination is measured from the celestial equator.

TERRESTRIAL EQUATOR the great circle of the earth 90° from the poles or ends of the axis of rotation. Latitude is measured from the equator.

Equinox, one of the two points where the ecliptic intersects the celestial equator. Also the time when the sun is at this point.

AUTUMNAL EQUINOX the equinox which the sun reaches in autumn. Also the time when the sun is at that point, September 23.

VERNAL EQUINOX the equinox which the sun reaches in spring. This point is called the First point of Aries, since that sign of the zodiac begins with this point, the sign extending eastward from it 30°. The celestial meridian (see Colure) passing through this point is the zero meridian of the celestial sphere, from which celestial longitude is reckoned. The vernal equinox is also the time when the sun is at this point, about March 21, the beginning of the astronomical year. See Year.

Geocentric (je ō sen´ trik; from *ge*, earth; *centrum*, center),

THEORY of the solar system assumes the earth to be at the center of the solar system; see Ptolemaic system.

LATITIDE, see Latitude.

PARALLAX see Parallax.

Geodesy (je od´ e sy), a branch of mathematics or surveying which is applied to the determination, measuring, and mapping of lines or areas on the surface of the earth.

Gravitation, the attractive force by which all particles of matter tend to approach one another.

Gravity, the resultant of *(a)* the earth's attraction for any portion of matter rotating with the earth and *(b)* the centrifugal force due to its rotation. The latter force *(b)* is so small that it is usually ignored and we commonly speak of gravity as the earth's attraction for an object. Gravity is still more accurately defined in the Appendix.

Heliocentric (he li ō sen´ trik; from *helios*, sun; *centrum*, center).

THEORY of the solar system assumes the sun to be at the center of the solar system; also called the Copernican system (see Copernican system).

PARALLAX see Parallax.

Horizon (hō ri´ zon), the great circle of the celestial sphere cut by a plane passing through the eye of the observer at right angles to the plumb line.

DIP OF HORIZON. If the eye is above the surface, the curvature of the earth makes it possible to see beyond the true horizon. The angle formed, because of the curvature of the earth, between the true horizon and the visible horizon is called the *dip* of the horizon.

VISIBLE HORIZON, the place where the earth and sky seem to meet. At sea if the eye is near the surface of the water the true horizon and the visible horizon are the same, since water levels and forms a right angle to the plumb line.

Hour-circles, great circles of the celestial sphere extending from pole to pole, so called because they are usually drawn every 15° or one for each of the twenty-four hours of the day. While hour-circles correspond to meridians on the earth, celestial longitude (see Longitude) is not reckoned from them as they change with the rotation of the earth.

Latitude, when not otherwise qualified, geographical latitude is meant.

ASTRONOMICAL LATITUDE the distance in degrees between the plumb line at a given point on the earth and the plane of the equator.

CELESTIAL LATITUDE the distance in degrees between a celestial body and the ecliptic.

GEOCENTRIC LATITUDE the angle formed by a line from a given point on the earth to the center of the earth (nearly the same as the plumb line) and the plane of the equator.

GEOGRAPHICAL LATITUDE, the distance in degrees of a given point on the earth from the equator. Astronomical, geocentric, and geographical latitude are nearly the same (see discussion of Latitude in Appendix).

Local time, see Time.

Longitude.

CELESTIAL LONGITUDE the distance in degrees of a celestial body from lines passing through the poles of the ecliptic (see Ecliptic), called ecliptic meridians; the zero meridian, from which celestial longitude is reckoned, is the one passing through the First point of Aries (see Equinox).

TERRESTRIAL LONGITUDE, the distance in degrees of a point on the earth from some meridian, called the prime meridian.

Mass, the amount of matter in a body, regardless of its volume or size.

Mean solar time, see Time.

Meridian.

CELESTIAL MERIDIAN a great circle of the celestial sphere passing through the celestial poles and the zenith of the observer. The celestial meridian passing through the zenith of a given place constantly changes with the rotation of the earth.

TERRESTRIAL MERIDIAN, an imaginary line on the earth passing from pole to pole. A meridian circle is a great circle passing through the poles.

Month.

CALENDER MONTH the time elapsing from a given day of one month to the same numbered day of the next month; *e.g.*, January 3 to February 3. This is the civil or legal month.

SIDEREAL MONTH the time it takes the moon to revolve about the earth in relation to the stars; one exact revolution of the moon about the earth; it varies about three hours in length but averages 27.32166 d.

SYNODIC MONTH the time between two successive new moons or full moons. This is what is commonly meant by the lunar month, reckoned from new moon to new moon; its length varies about thirteen hours but averages 29.53059 d. There are several other kinds of lunar months important in astronomical calculations.

SOLAR MONTH, the time occupied by the sun in passing through a sign of the zodiac; mean length, 30.4368 d.

Nadir (nā´ der), the point of the celestial sphere directly under the place on which one stands; the point 180° from the zenith.

Neap tides, see Tides.

Nutation, a small periodic elliptical motion of the earth's axis, due principally to the fact that the plane of the moon's orbit is not the same as the plane of the ecliptic, so that when the moon is on one side of the plane of the ecliptic there is a tilting tendency given the bulging equatorial region. The inclination of the earth's axis, or the obliquity of the ecliptic, is thus slightly changed through a period of 18.6 years, varying each year from 0´´ to 2´´. (See Motions of the Axis in the Appendix.)

Oblateness, the same as ellipticity; see Ellipse.

Oblate spheroid, see Spheroid.

Obliquity (ob lik´ wi ty), of the ecliptic, see Ecliptic.

Opposition, see Syzygy.

Orbit, the path described by a heavenly body in its revolution about another heavenly body.

Parallax, the apparent displacement, or difference of position, of an object as seen from two different stations or points of view.

ANNUAL OR HELIOCENTRIC PARALLAX of a star is the difference in the star's direction as seen from the earth and from the sun. The base of the triangle thus formed is based upon half the major axis of the earth's orbit.

DIURNAL OR GEOCENTRIC PARALLAX of the sun, moon, or a planet is the difference in its direction as seen from the observers' station and the center of the earth. The base of the triangle thus formed is half the diameter of the equator.

Perigee (per. i je), the point in the orbit of the moon which is nearest to the earth. The term is sometimes applied to the nearest point of a planet's orbit.

Perihelion (per i he´ li on), the point in a planet's orbit which is nearest to the sun.

Poles.

CELESTIAL, the two points of the celestial sphere which coincide with the earth's axis produced, and about which the celestial sphere appears to rotate.

OF THE ECLIPTIC, the two points of the celestial sphere which are 90° from the ecliptic.

TERRESTRIAL, the ends of the earth's axis.

Ptolemaic system (tol e mā´ ik), the theory of the solar system advanced by Claudius Ptolemy (100–170 a.d.) that the earth is the center of the universe, the heavenly bodies daily circling around it at different rates. Called also the geocentric theory (see Geocentric).

Radius (plural, radii, rā´ di i), half of a diameter.

Radius Vector, a line from the focus of an ellipse to a point in the boundary line. Thus a line from the sun to any planet is a radius vector of the planet's orbit.

Refraction of light, in general, the change in direction of a ray of light when it enters obliquely a medium of different density. As used in astronomy and in this work, refraction is the change in direction of a ray of light from a celestial body as it enters the atmosphere and passes to the eye of the observer. The effect is to cause it to seem higher than it really is, the amount varying with the altitude, being zero at the zenith and about 36´ at the horizon.

Revolution, the motion of a planet in its orbit about the sun, or of a satellite about its planet.

Rotation, the motion of a body on its axis.

Satellite, a moon.

Sidereal day, see Day.

Sidereal year, see Year.

Sidereal month, see Month.

Sidereal time, see Time.

Signs of the zodiac, its division of 30° each, beginning with the vernal equinox or First point of Aries.

Solar times, see Time.

Solstices (sol´ stis es; *sol,* sun; *stare*, to stand), the points in the ecliptic farthest from the celestial equator, also the dates when the sun is at these points; June 21, the summer solstice; December 22, the winter solstice.

Spheroid (sfe´ roid), a body nearly spherical in form, usually referring to the mathematical form produced by rotating an ellipse about one of its axes; called also an ellipsoid or spheroid of revolution (in this book, a spheroid of rotation).

OBLATE SPHEROID, a mathematical solid produced by rotating an ellipse on its minor axis (see Ellipse).

PROLATE SPHERIOD a mathematical solid produced by rotating an ellipse on its major axis (see Ellipse).

Syzygy (siz′ i jy; plural, syzygies), the point of the orbit of the moon (planet or comet) nearest to the earth or farthest from it. When in the syzygy nearest the earth, the moon (planet or comet) is said to be in conjunction; when in the syzygy farthest from the earth it is said to be in opposition.

Time.

APPRARENT SOLAR TIME, the time according to the actual position of the sun, so that twelve o'clock is the moment when the sun's center passes the meridian of the place (see Day, apparent solar).

ASTRONOMICAL TIME, the mean solar time reckoned by hours numbered up to twenty-four, beginning with mean solar noon (see Day, astronomical).

CIVIL TIME legally accepted time; usually the same as astronomical time except that it is reckoned from midnight. It is commonly numbered in two series of twelve hours each day, from midnight and from noon, and is based upon a meridian prescribed by law or accepted as legal (see Day, civil).

EQUATION OF TIME, see Equation of time.

SIDEREAL TIMEthe time as determined from the apparent rotation of the celestial sphere and reckoned from the passage of the vernal equinox over a given place. It is reckoned in sidereal days (see Day, sidereal).

SOLAR TIMEis either apparent solar time or mean solar time, reckoned from the mean or average position of the sun (see Day, solar day).

SYANDARD TIME, the civil time that is adopted, either by law or usage, in any given region; thus practically all of the people of the United States use time which is five, six, seven, or eight hours earlier than mean Greenwich time, being based upon the mean solar time of 75°, 90°, 105°, or 120° west of Greenwich.

Tropical year, see Year.

Tropics.

ASTRONOMICAL the two small circles of the celestial sphere parallel to the celestial equator and 23°27′ from it, marking the northward and southward limits of the sun's center in its annual (apparent) journey in the ecliptic; the northern one is called the tropic of Cancer and the southern one the tropic of Capricorn, from the signs of the zodiac in which the sun is when it reaches the tropics.

GEOGRAPHAL, the two parallels corresponding to the astronomical tropics, and called by the same names.

Vernal equinox, see Equinox, vernal.

Year.

ANOMALISTIC YEAR (a nom a lis´ tik), the time of the earth's revolution from perihelion to perihelion again; length 365 d., 6 h., 13 m., 48 s.

CIVIL YEAR the year adopted by law, reckoned by all Christian countries to begin January 1st. The civil year adopted by Protestants and Roman Catholics is almost exactly the true length of the tropical year, 365.2422 d., and that adopted by Greek Catholics is 365.25 d. The civil year of non-Christian countries varies as to time of beginning and length, thus the Turkish civil year has 354 d.

LUNAR YEAR the period of twelve lunar synodical months (twelve new moons); length, 354 d.

SIDEREAL YEAR, the time of the earth's revolution around the sun in relation to a star; one exact revolution about the sun; length, 365.2564 d.

TROPICAL YEAR the period occupied by the sun in passing from one tropic or one equinox to the same again, having a mean length of 365 d. 5 h. 48 m. 45.51 s. or 365.2422 d. A tropical year is shorter than a sidereal year because of the precession of the equinoxes.

Zenith (ze´ nith), the point of the celestial sphere directly overhead; 180° from the nadir.

Zodiac (zō´ di ak), an imaginary belt of the celestial sphere extending about eight degrees on each side of the ecliptic. It is divided into twelve equal parts (30° each) called signs, each sign being somewhat to the west of a constellation of the same name. The ecliptic being the central line of the zodiac, the sun is always in the center of it, apparently traveling eastward through it, about a month in each sign. The moon being only about 5° from the ecliptic is always in the zodiac, traveling eastward through its signs about 13° a day.

INDEX

D

H

I

J

K

L

M

T